SpringerBriefs in Applied Sciences and Technology

PoliMI SpringerBriefs

More information about this subseries at http://www.springer.com/series/11159
http://www.polimi.it

Shima Zahmatkesh · Emanuele Della Valle

Relevant Query Answering over Streaming and Distributed Data

A Study for RDF Streams and Evolving Web Data

MILANO 1863

Shima Zahmatkesh 🅳
DEIB
Politecnico di Milano
Milano, Italy

Emanuele Della Valle 🅳
DEIB
Politecnico di Milano
Milano, Italy

ISSN 2191-530X ISSN 2191-5318 (electronic)
SpringerBriefs in Applied Sciences and Technology
ISSN 2282-2577 ISSN 2282-2585 (electronic)
PoliMI SpringerBriefs
ISBN 978-3-030-38338-1 ISBN 978-3-030-38339-8 (eBook)
https://doi.org/10.1007/978-3-030-38339-8

This Springer imprint is published by the registered company Springer Nature Switzerland AG
The registered company address is: Gewerbestrasse 11, 6330 Cham, Switzerland

Preface

Nowadays, Web applications have often to combine highly dynamic data streams with data distributed over the Web to provide relevant answers for their users. Social Media analysis and Web of Things are contexts that require this type of Web applications. Social Media analysis often needs to look up the profiles of influencers most mentioned in a stream of posts. Web of Thing applications often have to join streams of sensor observations with data about the platforms that host the sensors to extract samples that best represent a phenomenon. In those settings, responding in a timely fashion, i.e., being reactive, is one of the most important requirements. However, when trying to join data streams with distributed data on Web, the time to access and fetch the distributed data can be so high that applications are at risk of losing reactiveness.

In particular, this book focuses on RDF Stream Processing (RSP) engines because they offer a query language (namely, RSP-QL) that eases the development of this type of Web queries and caching features that keep RSP engines reactive if the distributed data is static. However, RSP engines are also at risk of losing reactiveness when the distributed data is evolving.

For this reason, this book introduces the ACQUA framework to address the problem of evaluating RSP-QL queries over streaming and evolving distributed data. ACQUA keeps a local replica of the distributed data and offers an expertly defined maintenance process to refresh the replica over time. The users of ACQUA can set a refresh budget to control the number of elements to refresh in the replica before each evaluation. If set correctly, the refresh budget guarantees by construction that the RSP engine is reactive. When the maintenance process has enough budget to refresh all the stale elements, the answers of the RSP engine are exact. Otherwise, the maintenance process tries to approximate the result. Notably, the maintenance process is designed to gracefully decrease the accuracy of the answer when the refresh budget diminishes.

The remainder of the book presents extensions of the ACQUA framework for relevant query answering. It first introduces ACQUA.F to reactively answer to queries that pose a filter condition on the distributed data. For instance, a social media analysis may ask for users with more than one million followers that are mentioned in the last five minutes. Then, it brings in rank aggregation as a way to combine the maintenance processes proposed in ACQUA and those proposed in ACQUA.F.

Finally, the book focuses on continuous top-k queries and introduces AcquaTop. Consider, for instance, a mobile application for supporting people in parking in a crowded city. An RDF stream continuously reports the positions of the cars looking for a parking lot, while a Web service returns the number of free parking lots per city district. The continuous top-k query has to return the areas (around the car of the user's mobile App) where there is the largest number of free parking lots and the smallest number of cars looking for parking.

The authors of this book thank Dr. Daniele Dell'Aglio for his comments and support during the Ph.D. studies of Dr. Shima Zahmatkesh. They also thank Dr. Soheila Dehghanzadeh for reviewing Chap. 3. Last but not least, they acknowledge the contribution of Prof. Abraham Bernstein, Dr. Alessandra Mileo, and Dr. Shen Gao in shaping the ACQUA framework.

Milan, Italy   Shima Zahmatkesh
November 2019   Emanuele Della Valle

Contents

Chapter 1
Introduction

Abstract This chapter offers a brief introduction to continuous query evaluation over streaming and distributed data. The motivations behind the research question are illustrated through some examples from Social Media and industrial IoT. They explain the need of combining streaming data with distributed data on the Web in order to answer queries. The aim of this study is to overcome the problem of responding to the queries in a timely fashion in the proposed setting. This research considers different classes of queries and proposes a framework and various algorithms to generate relevant results. This chapter describes the main contributions of the work and focuses on the research questions and the approach followed during the research study. Finally, the layout of the book is presented.

Keywords Relevant query answering · Streaming data · Evolving distributed data

1.1 Motivation

Many modern applications require to combine highly dynamic data streams with distributed data, which slowly evolves, to continuously answer queries in a reactive way.[1] Consider the following two examples in Social Media and industrial IoT.

In social content marketing, advertisement agencies may want to continuously detect emerging influential Social Network users in order to ask them to endorse their commercials. To do so, they monitor users' mentions and the number of followers in micro-posts across Social Networks. They need to select those users who are not already known to be an influencer, are highly mentioned, and whose number of followers grows fast. It is worth to note that (1) users' mentions continuously arrive, (2) the number of followers may change in seconds, and (3) agencies have around a minute to detect them. Otherwise, the competitors may reach the emerging influencer

---

[1] A program is reactive if it maintains a continuous interaction with its environment, but at a speed which is determined by the environment, not by the program itself [1]. Real-time programs are reactive, but reactive programs can be non real-time as far as they provide result in time to successfully interact with the environment.

1

S. Zahmatkesh and E. Della Valle, *Relevant Query Answering over Streaming and Distributed Data*, PoliMI SpringerBriefs,
https://doi.org/10.1007/978-3-030-38339-8_1

sooner than them or the attention to the emerging influencer may drop. It is possible to formulate this information need as a continuous query of the form:

Example 1 Return every minute the users who are not well-known influencer, are mentioned the most and whose number of followers is growing the fastest.

Twitter API can be used to answer the query above, In particular, a streaming API[2] can provide access to the sample stream of micro-posts. On average, some 2,000 account mentions can be obtained per minute. The sample stream contains around 1% of the tweets. Therefore, *at scale* using the API that streams all the tweets, around 200,000 mentions per minute would be found. To obtain the number of followers of each mentioned account, the streaming APIs cannot be used and the REST service[3] is needed. This service returns fully-hydrated user descriptions for up to 100 users per request, thus 2,000 requests per minute should return us the information needed to answer the query. Unfortunately, this naïve approach will fail to be reactive for at least two reasons.

First of all, as it often happens on the Web, the service is rate limited to 300 requests every 15 min, i.e., 20 requests per minute, and its terms of usage forbids parallel requests. Notably, such a rate limit prevents to answer the query at scale while being reactive. It is at most enough to gather the number of followers of users mentioned in the sample stream.

Secondly, even if the REST service would not be rate limited, each request takes around 0.1s. Therefore, in one minute, at most 600 requests can be asked, which, again, is not enough to answer the query in a timely fashion.

Here is one more example, this time it is about manufacturing companies that use automation and instrumented their production lines with sensors, i.e., a typical scenario for the Web of Things[4] (WoT). In this setting, a production line consists of various machineries using different tools. For each instrument used by each machinery in the production line, the companies keep static data such as brand, type, installation date, etc. In addition, they also track the usage of each instrument mounted on each machine for maintenance purposes. A machine can automatically change the instrument it uses every minute. The information about when an instrument is in use on a machine and when it was last maintained is typically stored in an Enterprise Resource Planning (ERP) system that is not in the production site. The sensors in those companies track the environmental conditions of all machineries. They continuously stream several thousands of variables per second per production line, e.g., temperature, pressure, vibration, etc of each machine. They normally stream all those information. A common usage of all this information is the reactive detection of the environmental condition that can affect the quality of the products. For example, to check (directly on the production site) the effects of vibration on the quality of the product, it is possible to formulate a continuous query such as:

---

[2]https://dev.twitter.com/streaming/reference/get/statuses/sample.

[3]https://dev.twitter.com/rest/reference/get/users/lookup.

[4]https://www.w3.org/WoT/.

Example 2 Return every minute the list of products made with instruments that are the least recently maintained and are mounted on machines that show the highest vibrations.

As in the Social Media scenario, answering this query in a reactive manner is challenging since it joins thousands of observations streamed per seconds with the information stored in the ERP. If, as it is often the case, the network between the production line and the ERP has a latency of 100 ms, it may be impossible to perform the entire join.

RDF Stream Processing (RSP) community has recently started addressing the problem of evaluating queries over streaming and distributed data. RSP engine is an adequate framework to develop this type of queries [3]. The query has to use federated SPARQL syntax[5] which is supported by different RSP query languages (RSP-QL [5]). For instance, a simplified version of Example 1 can be formulated in the following continuous RSP-QL query:

Return every minute the top 3 most popular users who are most mentioned on Social Networks in the last 10 min.

Listing 1.1 shows how the query can be encoded as a continuous top-k query using the syntax proposed in [4]. At each query evaluation, the WHERE clause at Lines 4–7 are matched against the data in a window :W which opens on the stream of micro-posts and in the remote SPARQL service :BKG that contains the number of followers. Function F computes the score as a weighted sum of the inputs normalized in [0..1]. The users are ordered by their scores, and the number of results is limited to 3.

```
1  REGISTER STREAM :TopkUsersToContact AS
2  SELECT ?user F(?mentionCount,?followerCount) AS ?score
3  FROM NAMED WINDOW :W ON :S [RANGE 10m STEP 1m]
4  WHERE{
5        WINDOW :W {?user :hasMentions ?mentionCount}
6        SERVICE :BKG {?user :hasFollowers ?followerCount }
7  }
8  ORDER BY DESC (?score)
9  LIMIT 3
```

Listing 1.1 Sketch of the query studied in the problem

1.2 Problem Statement and Research Question

As stated before, in continuous query answering, being reactive and responding in a timely fashion is one of the most important requirements. However, when trying to join data streams with distributed data on the Web, the time to access and fetch the distributed data can be so high that applications may lose their reactiveness. Although

---

[5]http://www.w3.org/TR/sparql11-federated-query/.

RSP engines are suitable for posing this type of queries, they are also at risk of losing reactiveness when accessing distributed data over the Web.

RSP engines can remain reactive using a local replica of the distributed data, and offer a maintenance process to refresh it over time based on the refresh budget. However, if the refresh budget is not enough to refresh all data in the replica, some elements become stale and the query evaluation is no longer correct.

This, in general, may be unacceptable, but in some cases, as in the examples above, approximated results may be acceptable. This is especially true if the user can control the relevancy of results by ordering them. In this setting the latency and the high relevancy of the first results are essential, the completeness has little importance, and approximation for less relevant results is acceptable. Resource consumption is another important metric in the problem space, because the solution has to scale to thousands of concurrent users as current search engines do.

In order to attack the problem of query evaluation over streaming and evolving distributed datasets, this study answer the following question:

RQ. *Is it possible to optimize query evaluation in order to continuously obtain the most relevant combinations of streaming and evolving distributed data, while guaranteeing the reactiveness of the engine?*

The goal of this study is to continuously answer queries that require to (*i*) find the most relevant answers, (*ii*) join data streams with slowly evolving datasets published on the Web of Data, and (*iii*) respect the reactiveness constraints imposed by the users.

To answer the research question, this study, first, focuses on RSP-QL queries that contains WINDOW and SERVICE clauses, which join a data stream with a distributed dataset, and, later, considers more complicated queries. The class of queries which are studied in this work are the following:

- Queries that contain WINDOW and SERVICE clauses.
- Queries that contain a FILTER clause inside the SERVICE clause. Exploiting the presence of a Filtering Threshold, only a subset of the mappings are returned by the SERVICE clause.
- Top-k queries that return the top-k most relevant results to the user based on a predefined scoring function that combines variables appearing in the WINDOW and SERVICE clauses.

1.3   Approach and Contributions

This study focuses on the RSP engine context and proposes a framework named Approximate Continuous Query Answering over streaming and distributed data (shortly, ACQUA) [2]. The baseline version of ACQUA proposes to compute the answer of the SERVICE clause at query registration time and to store the result mappings in a local replica. Then, given that the distributed data evolves (e.g. the number of followers in Example 1), maintenance policies maximize the freshness of the mappings in the replica, while guaranteeing the reactiveness of the engine. ACQUA includes both traditional policies such as random policy (RND) and Least-Recently Used policy (LRU) as well as policies specifically tailored to the streaming case such as Window Based Maintenance policy (WBM). ACQUA framework focuses on the class of queries that contains WINDOW and SERVICE clauses, and assumes a 1:1 relationship between the streaming and the distributed data.

The first extension of ACQUA (named Filter Update Policy) considers the class of queries that contains a FILTER in the SERVICE clause [8]. This is a rough way to express relevance. The Filter Update Policy exploits the following intuition: when spending the budget to check the freshness of data in the replica, it is better to focus on data items that are likely to pass the filter condition and may affect the future evaluation.

Then, the Filter Update Policy is combined with ACQUA policies. The results are WBM.F, LRU.F, and RND.F policies, collectively named ACQUA.F policies. In the proposed algorithm, Filter Update Policy and one of the ACQUA policies are applied in a pipe, assuming that it is simple to determine a priori the *band* around filtering condition to focus on.

Relaxing such an assumption on the *band* in the previous approach, the Rank Aggregation Policies are proposed, which let each policy to rank data items according to its criterion (i.e., to express its opinion), and then, aggregates them to take into account all opinions [6]. In the rank aggregation approach, three algorithms are proposed, which combine Filter Update Policy with ACQUA policies, respectively, named LRU.F$^+$, WBM.F$^+$, and WBM.F* (improved version of WBM.F$^+$) [10].

The second extension of ACQUA focuses on the class of continuous top-k queries [9]. It starts from the state-of-the-art approach for continues top-k query evaluation [7], and proposes a solution that also works when joining the stream with evolving distributed dataset. Considering the architectural approach presented in ACQUA as a guideline, the AcquaTop framework is proposed. In order to approximate as much as possible the correct answer, *two maintenance policies* (AT-BSM, and AT-TSM) are proposed for updating the local replica. They are specifically tailored to top-k query answering.

1.4   Structure of the Book

The book is structured as follows:

- Chapter 2 defines the relevant background concepts on Semantic Web, RDF stream processing, top-k query answering, and reviews the state of the art in top-k query monitoring over streaming data.

- Chapter 3 introduces **ACQUA** as a framework for the approximate query evaluation over streaming and distributed data focusing on 1:1 join relationship between streaming and distributed data. This chapter is based on the work of Dehghanzadeh et al. [2].
- Chapter 4 introduces **ACQUA.F** for the approximate query evaluation over streaming and distributed data focusing on the class of queries that contains a FILTER clause. In this chapter, assuming that determining a priori a band around the filter threshold is simple, various maintenance policies are presented in order to keep the local replica fresh. The experimental evidence shows that those policies outperform the state-of-the-art ones. This chapter is based on the work of Zahmatkesh et al. [8].
- Chapter 5 further investigates the approximate query evaluation over streaming and distributed data focusing on combining various maintenance policies presented in the previous chapter. It shows that rank aggregation allows relaxing the assumption of knowing a priori the band to focus on. The experimental evaluations show that relaxing the assumption, the same or even better performance can be achieved. This chapter is based on the work of Zahmatkesh et al. [10].
- Chapter 6 presents **AcquaTop** framework and the result of the investigations on top-k query answering over streaming and evolving distributed data. This chapter is based on the work of Zahmatkesh et al. [9].
- Chapter 7 concludes with a review of the contributions, a discussion of the limits, and presenting directions to future works.

References

1. Berry G (1989) Real time programming: special purpose or general purpose languages. PhD thesis, INRIA
2. Dehghanzadeh S, Dell'Aglio D, Gao S, Della Valle E, Mileo A, Bernstein A (2015) Approximate continuous query answering over streams and dynamic linked data sets. In: 15th international conference on web engineering, Switzerland, Jun 2015
3. Della Valle E, Dell'Aglio D, Margara A (2016) Taming velocity and variety simultaneously in big data with stream reasoning: tutorial. In: Proceedings of the 10th ACM international conference on distributed and event-based systems. ACM, pp 394–401
4. Dell'Aglio D, Calbimonte J-P, Della Valle E, Corcho O (2015) Towards a unified language for rdf stream query processing. In: International semantic web conference. Springer, pp 353–363
5. Dell'Aglio D, Della Valle E, Calbimonte J-P, Corcho Ó (2014) RSP-QL semantics: a unifying query model to explain heterogeneity of RDF stream processing systems. Int J Semant Web Inf Syst 10(4):17–44
6. Dwork C, Kumar R, Naor M, Sivakumar D (2001) Rank aggregation methods for the web. In: WWW. ACM, pp 613–622
7. Yang D, Shastri A, Rundensteiner EA, Ward MO (2011) An optimal strategy for monitoring top-k queries in streaming windows. In: Proceedings of the 14th international conference on extending database technology. ACM, pp 57–68
8. Zahmatkesh S, Della Valle E, Dell'Aglio D (2016) When a filter makes the difference in continuously answering sparql queries on streaming and quasi-static linked data. In: International conference on web engineering. Springer, pp 299–316

9. Zahmatkesh S, Della Valle E, Continuous top-k approximated join of streaming and evolving distributed data. Semantic Web. (In press). Available online http://semantic-web-journal.net/content/continuous-top-k-approximated-join-streaming-and-evolving-distributed-data-0#

10. Zahmatkesh S, Della Valle E, Dell Aglio D (2017) Using rank aggregation in continuously answering sparql queries on streaming and quasi-static linked data. In: Proceedings of the 11th ACM international conference on distributed and event-based systems. ACM, pp 170–179

Chapter 2
Background

Abstract This chapter presents the preliminary contents needed for the rest of the book. The beginning of the chapter introduces the foundation concept about Semantic Web data and query model and a review of state-of-the-art works. Then, focusing on RDF Stream Processing, it presents the semantics of the RDF stream query language (RSP-QL), which is important for precisely formalize the problem in the following chapters. Then, the chapter proceeds to introduce Top-k Query Answering, and Top-k Query Monitoring over the Data Stream and the related works in these domains. Finally, it offers a review of the metrics used in this study to evaluate the accuracy and relevancy of the answers in the result.

Keywords RDF stream data · RDF stream query language · Top-k query answering

2.1 RDF Data Model and SPARQL Query Language

The Resource Description Framework (RDF) is a standard framework proposed by W3C used for representing data on the Web [37]. The main structure, which is known as a *triple*, consists of *subject*, *predicate*, and *object*. Let I, B and L be three pairwise disjoint sets, defined as the set of IRIs, blank nodes and literals, respectively. An RDF term is defined as an element of the set $I \cup B \cup L$.

Definition 2.1 (*RDF statement and RDF graph*) An RDF statement is a triple $(s, p, o) \in (I \cup B) \times (I) \times (I \cup B \cup L)$, while a set of RDF statements is called an RDF graph, which is a directed, labeled graph that represents Web resources.

For instance, Listing 2.1 shows the number of followers for different users in Social Media encoded as RDF statements:

```
1  :Alice :hasFollowers 6300 .
2  :Bob   :hasFollowers 5500 .
3  :Carol :hasFollowers 10800 .
4  :Eve   :hasFollowers 11100 .
```

Listing 2.1 Examples of RDF statements

The SPARQL Protocol and RDF Query Language (SPARQL) is a standard query language, which is able to retrieve and manipulate data stored in RDF format [27]. Notably, it includes a model (namely, SPARQL 1.1 Federated query extension) to direct a part of the query to a particular SPARQL endpoint [2]. A SPARQL query [24] is defined through a triple (E, DS, QF), where E is the algebraic expression, DS is the data set and QF is the query form.

A SPARQL query typically contains one or more triple patterns called a basic graph pattern as an algebraic expression E in the WHERE clause. Comparing to the RDF statement, triple patterns may contain variables in place of resources.

Definition 2.2 (*Graph Pattern*) In addition to I, B and L, let V be the set of variables (disjointed with the other sets); graph patterns expressions are recursively defined as follows:

- a basic graph pattern (i.e. set of triple patterns $(s, p, o) \in (I \cup B \cup V) \times (I \cup V) \times (I \cup B \cup L \cup V))$ is a graph pattern;
- let P_1 and P_2 be two graph patterns, P_1 *UNION* P_2, P_1 *JOIN* P_2, and P_1 *OPT* P_2 are graph patterns;
- let P be a graph pattern and F a built-in condition, P *FILTER* F is a graph pattern;
- let P be a graph pattern and $u \in (I \cup V)$, the expressions *SERVICE* u P, and *GRAPH* u P are graph patterns;

A SPARQL built-in condition consists of the elements of the set $(I \cup L \cup V)$ and constants, logical connectives $(\neg, \vee, \wedge)$, the binary equality symbol $(=)$, ordering symbols $(<, \leq, \geq, >)$, and unary predicates such as *bound*, *isBlank*, *isIRI*.

Definition 2.3 Let P be a graph pattern, $var(P)$ denotes the set of variable names occurring in P $(var(P) \sqsubseteq V)$.

SPARQL dataset DS defines as a set of pairs of symbols and graphs associated with those symbols, i.e., $DS = \{(def, G), (g_1, G_1), \ldots, (g_k, G_k)\}$ with $k \geq 0$, where the default graph G is identified by the special symbol $def \notin I$ and the remaining ones are named graphs (G_i) and are identified by IRIs $(g_i \in I)$. Function $names(DS)$ given a dataset DS returns the set of graph names $\{g_1, \ldots, g_k\}$.

The SPARQL language specifies four different query form QF for different purposes: SELECT, CONSTRUCT, ASK, and DESCRIBE.

In order to define the semantics of SPARQL evaluation, first, some definitions from [25] are introduced, focusing on the minimum information required to understand the book.

Definition 2.4 (*Solution Mapping*) The evaluation of graph pattern expressions produces sets of solution mappings. A solution mapping is a function that maps variables to RDF terms, i.e., $\mu : V \rightarrow (I \cup B \cup L)$. $dom(\mu)$ denotes the subset of V where μ is defined. $\mu(x)$ indicates the RDF term resulting by applying the solution mapping to variable x.

Definition 2.5 (*Compatible Solution Mappings*) Two solution mappings μ_1 and μ_2 are *compatible* ($\mu_1 \sim \mu_2$) if the two mappings assign the same value to each variable in $dom(\mu_1) \cap dom(\mu_2)$, i.e., $\forall x \in dom(\mu_1) \cap dom(\mu_2)$, $\mu_1(x) = \mu_2(x)$.

Definition 2.6 (*Join Operator*) Let Ω_1 and Ω_2 be two sets of solution mappings, the join is defined as:

$$\Omega_1 \bowtie \Omega_2 = \{\mu_1 \cup \mu_2 | \mu_1 \in \Omega_1, \mu_2 \in \Omega_2, \mu_1 \sim \mu_2\} \tag{2.1}$$

Definition 2.7 (*Filter Operator*) Let Ω be a set of solution mappings, and $expr$ be an expression. The Filter is defined as:

$$Filter(expr, \Omega) = \{\mu | \mu \in \Omega, \text{ and } expr(\mu) \text{ is an expression that has an effective boolean value of true.}\} \tag{2.2}$$

Definition 2.8 (*Order By Operator*) Let Ψ be a sequence of solution mappings. Order By is defined as:

$$OrderBy(\Psi, condition) = \{\mu | \mu \in \Psi \text{ and the sequence satisfies the ordering condition}\} \tag{2.3}$$

Definition 2.9 (*Slice Operator*) Let Ψ be a sequence of solution mappings. Slice is defined as:

$$Slice(\Psi, start, length)[i] = \Psi[start + i] \text{ for } i = 0 \text{ to } (length - 1) \tag{2.4}$$

The evaluation of SPARQL query is represented as a set of solution mappings. The SPARQL evaluation semantics of an algebraic expression E is denoted as $[\![E]\!]_G^D$, where D is the dataset with active graph G. The function gets the algebraic expression E, and returns a set of mappings.

Listing 2.2 shows an example of SPARQL query that asks for the users whose number of followers is above 100,000. Considering the data in Listing 2.1, the result of the query will be {Carol, Eve}.

```
1  SELECT ?user
2  WHERE {
3       ?user :hasFollowers ?followerCount
4       FILTER (?followersCount > 10000)
5  }
```

Listing 2.2 Example of SPARQL query

Definition 2.10 Lets D be an RDF dataset, t a triple pattern, P, P_1, and P_2 graph patterns, and F a build-in condition. The evaluation of basic graph pattern, JOIN, and FILTER are defined as follow:

$$[\![t]\!]_G^D = \{\mu | dom(\mu) = var(t) \text{ and } \mu(t) \in D\} \tag{2.5}$$

$$[\![P_1 \; JOIN \; P_2]\!]_G^D = [\![P_1]\!]_G^D \bowtie [\![P_2]\!]_G^D \qquad\qquad (2.6)$$

$$[\![P \; FILTER \; F]\!]_G^D = \{\mu | \mu \in [\![P]\!]_G^D \text{ and } \mu \text{ satisfies } F\} \qquad (2.7)$$

In the federated SPARQL, SERVICE operator is defined to specify the IRI of the SPARQL endpoint where the related expression will be executed. The evaluation of the SERVICE operator is defined as follows [6]:

Definition 2.11 (*SERVICE evaluation*) For evaluation of SERVICE operator, lets graph pattern $P = SERVICE \; c \; P_1$, the evaluation of graph P over dataset D, and the active graph G defines as:

$$[\![P]\!]_G^D = \begin{cases} [\![P_1]\!]_{graph(def,ep(c))}^{ep(c)} & \text{if } c \in dom(ep) \\ \{\mu_0\} & \text{if } c \in I \setminus dom(ep) \\ \{\mu \cup \mu_c \mid \exists s \in dom(ep) : \mu_c = [c \rightarrow s], & \text{if } c \in V \\ \mu \in [\![P_1]\!]_{graph(def,ep(s))}^{ep(s)}\} \text{ and } \mu_c \sim \mu\} \end{cases} \qquad (2.8)$$

where $c \in I$, and ep is a partial function from the set I of IRIs, and for every $c \in I$, if $ep(c)$ is defined, then $ep(c) = D_c$, which is the own dataset of the SPARQL endpoint.

Based on this definition, if c is the IRI of SPARQL endpoint, the evaluation of the SERVICE clause, is equal to the evaluation of the graph pattern P_1 in the SPARQL endpoint specified by c. But, if c is not the IRI of SPARQL endpoint, the query cannot be evaluated and the variables in P_1 leaves unbounded.

Finally, if $c \in V$, the pattern *SERVICE ?X P* is defined by assigning all the values s in the domain of function ep to variable $?X$. The semantics of evaluation pattern *SERVICE ?X P* requires the evaluation of P over every SPARQL endpoints, which is infeasible unless the variable $?X$ is bound to a finite set of IRIs. Buil-Aranda et al. [6] provide a formalization for this concept as follows:

Definition 2.12 (*Boundedness*) Let P be a SPARQL query and $?X \in var(P)$. Then $?X$ is bound in P if one of the following conditions holds:

- P is either a graph pattern or a VALUES query, and for every dataset DS, every RDF graph G in DS, and every $\mu \in [\![P]\!]_G^{DS} : ?X \in dom(\mu)$ and $\mu(?X) \in (dom(DS) \cup names(DS) \cup dom(P))$.
- P is a SELECT query ($SELECT \; W \; P_1$) and $?X$ is bound in P_1.

2.2 Federated SPARQL Engines

There exist different engines that support the SPARQL 1.1 Federated Query extension such as ARQ, and SPARQL-DQP [6], or implement a distributed query processing like FedX [30], and DARQ [28].

ARQ[1] is a query engine contained in the Jena Framework that supports the SPARQL query language. The query processing in ARQ contains the following components: Parser, Algebra Generator, High-Level Optimizer, and Low-Level Optimizer.

Quilitz et al. [28] present DARQ, which is a federated SPARQL query engine. DARQ provides transparent query access to multiple SPARQL services, by adopting an architecture of mediator based information systems [35]. In DARQ engine, query processing consists of four stages: parsing, query planning, optimization, and query execution. For the parsing stage, the DARQ query engine reuses the parser of ARQ. Service descriptions describe the data available from an endpoint and allow the definition of limitations on access patterns. This information is used by the engine for query planning and optimization. In the query planning stage, the engine finds the relevant sources, and decomposes the query into sub-queries, according to the information in the service descriptions. Each sub-query can be answered by an individual data source. In the next stage, the query optimizer takes the sub-queries and generates a feasible and cost-effective query execution plan, using logical and physical query optimization. Finally, the plan is executed, and sub-queries are sent to the data sources and the results are integrated.

Schwarte et al. [30] propose join processing and grouping techniques to develop an efficient federated query processing. FedX is a practical solution for efficient federated query processing on Linked Data sources. FedX allows virtual integration of heterogeneous Linked Open Data sources and presents new join processing strategies that minimize the number of requests sent to the federated resources. The proposed Exclusive groups have a central role in the FedX optimizer that sends the triple patterns together as a conjunctive query to the endpoint instead of sending them sequentially, which minimize request number.

Buil–Aranda et al. [6] propose a federated SPARQL query engine named SPARQL-DQP, which supports SPARQL 1.1 Federated Query extension. They formalize the semantics of the SERVICE clause and introduce the definition of service-boundedness and service-safeness conditions.

They also provide static optimizations for queries that contain the OPTIONAL operator, using the notion of well-designed SPARQL graph patterns. These optimizations reduce the number of intermediate results being transferred and joined in federated queries.

They build SPARQL-DQP by extending a query federation system (OGSA-DQP [11]), on top of a data workflow system (OGSADAI [18]), which targets large amounts of data in e-Science applications. They consider SPARQL as a new query language and define RDF data sources as a new type of resource.

SPARQL-DQP accepts the SPARQL query, and the parser generates the abstract syntax tree. Then, the SPARQL-DQP Builder produces a logical query plan, following the semantics of SPARQL 1.1. The OGSA-DQP chain of optimizers and the rewriting rules based on well-designed patterns are applied on the basic logical query

---

[1]http://jena.apache.org/documentation/query/index.html.

plan. The node for execution is selected. Finally, the generated remote requests are executed, and the results are collected and integrated.

Acosta et al. [1] present ANAPSID, an adaptive query engine for SPARQL endpoints. ANAPSID extends the adaptive query processing features presented in [31] to adapt query execution based on the available data and run-time condition.

ANAPSID is based on the architecture of wrappers and mediators [36] to evaluate query federations of SPARQL endpoints. The mediators decompose queries into sub-queries that can be executed by the remote endpoints, and gather data retrieved from the endpoints. The wrappers create calls to the endpoints and get the answers from them. In addition, mediators maintain information about endpoints such as capabilities, ontologies used to describe the data, and statistics about the content and performance. They implement query rewriting techniques, decompose queries into sub-queries against the endpoints, and gather data retrieved from the endpoints.

Mediators are composed of the following components: (*i*) a Catalog that contains information of the endpoints, (*ii*) a Query Decomposer that generates sub-queries and selects endpoints that are capable of executing each sub-query, (*iii*) a Query Optimizer that generates bushy execution plans by combining sub-queries, and (*iv*) an Adaptive Query Engine that implements different physical operators which are able to detect blocked resource or bursty data traffic, and incrementally generate results when the data arrives.

2.3 RSP-QL Semantic

RDF Stream Processing (RSP) [10] extends the RDF data model and query model considering the temporal dimension of data and the continues evolution of query over time. In the following, the definitions of RSP-QL [9] are introduced.

An RSP-QL query is defined by a quadruple $\langle ET, SDS, SE, QF \rangle$, where ET is a sequence of evaluation time instants, SDS is an RSP-QL dataset, SE is an RSP-QL algebraic expression, and QF is a query form.

In order to define SDS, first, it is needed to introduce the concepts of time, RDF stream, and window over an RDF stream, which creates RDF graphs by extracting relevant portions of the stream.

Definition 2.13 (*Time*) The *time* T is an infinite, discrete, ordered sequence of time instants $(t_1, t_2, ...)$, where $t_i \in \mathbb{N}$.

Definition 2.14 (*RDF Stream*) An RDF stream S is a potentially unbounded sequence of timestamped data items (d_i, t_i):

$$S = (d_1, t_1), (d_2, t_2), \ldots, (d_n, t_n), \ldots, \tag{2.9}$$

where d_i is an RDF statement, $t_i \in T$ the associated time instant, and for each data item d_i, it holds $t_i \leq t_{i+1}$ (i.e., the time instants are non-decreasing).

Beside RDF streams, it is possible to have static or quasi-static data, which can be stored in RDF repositories or embedded in Web pages. For that data, the time dimension of SDS can be defined through the notions of time-varying and instantaneous graphs. The time-varying graph $\overline{G}$ is a function that maps time instants to RDF graphs and instantaneous graph $\overline{G}(t)$ is the value of the graph at a fixed time instant t.

Definition 2.15 (*Time-based Window*) A time-based window $W(S)$ is a set of RDF statements extracted from a stream S, and defined through opening and closing time instance (i.e., o, and c time instance) where $W(S) = \{d \mid (d, t) \in S, t \in (o, c]\}$.

Definition 2.16 (*Time-based Sliding Window*) A *time-based sliding window* operator $\mathbb{W}$ [9], depicted in Fig. 2.1, takes an RDF stream S as input and produces a time-varying graph $G_{\mathbb{W}}$. $\mathbb{W}$ is defined through three parameters: ω – its width –, β – its slide –, and t^0 – the time stamp on which $\mathbb{W}$ starts to operate.

Operator $\mathbb{W}$ generates a sequence of time-based windows. Given two consecutive windows W_i, W_j defined in $(o_i, c_i]$ and $(o_j, c_j]$, respectively, it holds: $o_i = t_0 + i * \omega, c_i - o_i = c_j - o_j = \omega$, and $o_j - o_i = \beta$. The sliding window could be count- or time-based [3].

Definition 2.17 (*Evaluation Time*) The *Evaluation Time* $ET \subseteq T$ is a sequence of time instants at which the evaluation occurs. It is not practical to give ET explicitly, so normally ET is derived from an evaluation policy. In the context of this study, all the time instants, at which a window closes, belong to ET. For other policies see [9].

Active windows are defined as all the windows that contain the evaluation time in their duration. *Current window* is the window that closes in the current evaluation time. Given current window W_{cur}, and next window W_{nxt} as two consecutive windows defined in $(o_{cur}, c_{cur}]$ and $(o_{nxt}, c_{nxt}]$, respectively, *current evaluation time* is defined as the closing time of current window, c_{cur}, and *next evaluation time* as the closing time of next window, c_{nxt}.

An RSP-QL dataset SDS is a set composed by one default time-varying graph $\overline{G}_0$, a set of n time-varying named graphs $\{(u_i, \overline{G}_i)\}$, where $u_i \in I$ is the name of the element; and a set of m named time-varying graphs obtained by the application of time-based sliding windows over $o \leq m$ streams, $(u_j, \mathbb{W}_j(S_k))\}$, where $j \in [1, m]$, and $k \in [1, o]$. It is possible to determine a set of instantaneous graphs and fixed windows for a fixed evaluation time, i.e. RDF graphs, and to use them as input data for the algebraic expression evaluation.

An algebraic expression SE is a streaming graph pattern which is the extension of a graph pattern expression defined by SPARQL. It is composed by operators mostly inspired by relational algebra, such as joins, unions and selections. In addition to the ones defined in SPARQL, RSP-QL adds a set of *streaming operators (RStream, IStream, and DStream), to transform the query result in an output stream. Considering the recursive definition of the graph pattern, Streaming graph pattern expressions are extended as follows:

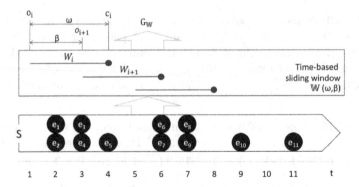

Fig. 2.1 The time-based sliding window operator dynamically selects a finite portion of the stream

- let P be a SPARQL graph pattern (see Definition 2.2) and $u \in (I \cup V)$, the expression $WINDOW\ u\ P$ is a graph pattern;
- let P be a graph pattern, $RStream\ P$, $IStream\ P$ and $DStream\ P$ are streaming graph patterns.

RSP-QL query form QF is defined as in SPARQL. Evaluation of a streaming graph pattern produces a set of solution mappings; RSP-QL extends the SPARQL evaluation function by adding the evaluation time instant: let $[\![P]\!]^t_{SDS(\bar{G})}$ be the evaluation of the graph pattern P at time t having $\bar{G} \in SDS$ as active time-varying graph. For the sake of space, in the following, the evaluation of the operators used in the remaining of the work are presented. The evaluation of a BGP P is defined as:

$$[\![P]\!]^t_{SDS(\bar{G})} = [\![P]\!]_{SDS(\bar{G},t)} \tag{2.10}$$

where the right element of the formula is the SPARQL evaluation (see Definition 2.11) of P over $SDS(\bar{G}, t)$. Being a SPARQL evaluation, $SDS(\bar{G}, t)$ identifies an RDF graph: an instantaneous graph $\bar{G}(t)$ if $\bar{G}$ is a time-varying graph, a fixed window generated by $\mathbb{W}(S)$ at time t ($\mathbb{W}(S, t) = \bar{G}_{\mathbb{W}}(t)$) if $\bar{G}$ is a time-based sliding window. Evaluations of JOIN, FILTER and WINDOW[2] are defined as follows:

$$[\![P_1\ JOIN\ P_2]\!]^t_{SDS(\bar{G})} = [\![P_1]\!]^t_{SDS(\bar{G})} \bowtie [\![P_2]\!]^t_{SDS(\bar{G})} \tag{2.11}$$

$$[\![P\ FILTER\ F]\!]^t_{SDS(\bar{G})} = \{\mu | \mu \in [\![P]\!]^t_{SDS(\bar{G})} \text{ and } \mu \text{ satisfies } F\} \tag{2.12}$$

$$[\![WINDOW\ u\ P]\!]^t_{SDS(\bar{G})} = [\![P]\!]^t_{SDS(\mathbb{W})} \text{ such that } (u, \mathbb{W}) \in SDS \tag{2.13}$$

Finally, the evaluation of $SERVICE\ u\ P$ consists in submitting the graph pattern P to a SPARQL endpoint located at u and produces a set Ω_S with the resulting mappings.

---

[2]In the following, $u \in I$ is assumed.

2.4 Top-k Query Answering

The top-k query answering problem has been studied in different domains like database, Semantic Web, and stream processing. In many application domains, end-users are only interested in the most important (top-k) query answers in the potentially huge answer space [16].

Definition 2.18 (*Top-k Query*) Top-k query gets a user-specified scoring function, and provides only the top k query answers with the highest score based on the scoring function.

Definition 2.19 (*Scoring Function*) The scoring function $\mathcal{F}(p_1, p_2, ..., p_n)$ generates score for each result of the query by aggregating multiple predicates, where p_i is a scoring predicate.

Most of the top-k processing techniques assume that the scoring function $\mathcal{F}$ is monotonic, i.e., $\mathcal{F}(x_1, ..., x_n) \geq \mathcal{F}(y_1, ..., y_n)$ when $\forall i : x_i \geq y_i$.

The property of monotone scoring functions leads to efficient processing of top-k queries. When objects from various ranked lists are aggregated using a monotone scoring function, an upper bound of the score for unseen objects can be derived. This property, which is used in different top-k processing algorithms, guarantees early termination of top-k processing. This book treats only monotonic scoring functions.[3]

The Listing 2.3 shows an example of top-k query in which returns the 3 youngest users with the highest number of followers:

```
1   SELECT ?user F(?followerCount, ?age) AS ?score
2   WHERE{ ?user :hasFollowers ?followerCount .
3          ?user :hasAge ?age
4   }
5   ORDER BY DESC (?score)
6   LIMIT 3
```

Listing 2.3 Example of Top-k query

The naive algorithm for top-k query first materializes the score for each result in the answer, then it sorts the result by the score. Fagin in [12] was the first to introduce the idea to interleave sorting and query evaluation. The sorting is performed by a rank operator, while the rest of the evaluation consumes the result of the rank operator. *Fagin's Algorithm (FA)* often performs better than the naive algorithm. Later, Fagin et al. [13] introduce another algorithm named *Threshold Algorithm (TA)*, which is stronger than FA. TA algorithm assumes random access beside sorted access to the

---

[3]The evaluation of top-k queries with generic scoring function is not straightforward, as they cannot eliminate items which are not in the top-k result in early stage. Zhang et al. [41] address this problem by modeling top-k query as an optimization problem. There is another category of queries that do not have scoring function, called skyline queries. The skyline queries give a set of answers which are not dominated by any other answer. Various researches study the skyline queries in database community such as [4, 7, 23].

separated lists related to attributes. They provide an optimal algorithm for cases where random access is impossible or expensive.

No Random Access (NRA) algorithms assume that only sorted access is available for each list. Natsev et al. [22] introduce $J*$ algorithm which is an example of NRA algorithm. They address the problem of incremental joins of multiple ranked inputs. The proposed algorithm can support joins of ranked inputs based on user-defined join predicates, and multiple levels of joins that arise in nested views.

Ilyas et al. [14, 15] present the generation of top-k result based on join over relations. They propose a physical query operator to implement the rank-join algorithm, so the new operator can be used in practical query engines, and query optimizer can optimize the query execution plan containing the new integrated rank-join operator. In [17], the relational query optimizer is extended to apply the rank-join operator during query plan creation.

Li et al. [19] introduce RankSQL, which is a system that supports efficient top-k query evaluation in relational database systems. They extended the relational algebra and proposed *rank-relational* model considering ranking as a first-class construct. They also extend the query optimizer and propose dimensional enumeration algorithm to optimize top-k query.

Yi et al. [40] introduce an approach to incrementally maintain the materialized top-k views, which can improve query performance. In general, materialized top-k view is not self-maintainable, as due to the deletions and updates on the base table, tuples may leave the top-k view. To refill the view, the underlying top-k query needs to evaluate again over the base table. The idea is to consider top-k' view instead of top-k view, where k' is a parameter that can change between k and parameter $k_{max} \geq k$ to reduce the frequency of re-computation of top-k view which is an expensive operation.

Interested readers can refer to [16], which is a survey on top-k query processing techniques in relational databases.

Various works on top-k query answering are also available in the Semantic Web community [20, 32–34, 39].

Magliacane et al. [20] improve the performance of top-k SPARQL queries by extending SPARQL algebra and considering order as a first class citizen, and propose an incremental execution model for the SPARQL-RANK algebra. They introduce ARQ-RANK, a rank-aware SPARQL query engine that builds on the SPARQL-RANK algebra and utilizes state-of-the-art rank-aware query operators. Authors, also propose a rank-aware join algorithm optimized for native RDF stores.

Wagner et al. [33] study the top-k join problem in a Linked Data setting where different sources are accessible through URI lookups. They discussed how existing top-k join techniques can be adapted to the Linked Data context. They also provide two optimizations. First, they propose strategies that use knowledge about resource, and provide tighter score bounds which lead to earlier termination. Second, they introduce an aggressive technique for pruning partial query results that cannot contribute to the final top-k result.

Wagner et al. [32] introduce an approximate join top-k algorithm for the Web of data. They extend the PBRJ framework [29] with a novel probabilistic component to estimate the probability of a partial query binding. For a given partial query binding, they estimate its probability for contributing to the final top-k results, and discard partial bindings with low probability. In the proposed framework, all needed score statistics are learned via a pay-as-you-go paradigm at runtime.

Wang et al. [34] propose a graph-exploration-based method for top-k SPARQL queries evaluation on RDF graphs. Once an entity with a potentially high score is found, the graph-exploration method is employed to find the candidate's corresponding sub-graph matches. They also introduce the index MS-tree to efficiently evaluate top-k queries in RDF data. Based on an MS-tree, they propose an optimized upper-bound computation method to obtain a tight upper bound.

Yang et al. [39] propose STAR, which is a top-k knowledge graph search framework. First, they propose an approach to optimize star query processing, then, using effective query decomposition and star query processing, they introduce a query optimization for answering general graph queries.

2.5 Top-k Query Monitoring Over the Data Stream

Starting from the mid 2000s, various works addressed the problem of top-k query answering on data stream [21, 26, 38] (a.k.a. top-k query monitoring). The bottleneck that all approaches try to avoid is the recomputation of the top-k query from scratch when new data arrives from the stream. All authors introduce novel techniques for incremental query evaluation.

Yang et al. [38] propose an optimal solution regarding CPU and memory complexity. The Authors introduce *Minimal Top-K candidate set (MTK)*,[4] which is necessary and efficient for continuous top-k query evaluation. They introduce a compact representation for predicted top-k results, named *super-top-k list*. They also propose *MinTopk algorithm* based on MTK set and finally, prove the optimality of the proposed approach.

Going into the details of [38], let's consider a window of size w that slides every β. When an object arrives in the current window, it will also participate in all w/β future windows. Therefore, a subset of top-k result in the current window, which also participates in all future windows, has the potential to contribute to the top-k result in future windows. The objects in predicted top-k result constitute the *MTK set.*

In order to reach optimal CPU and memory complexity, they propose a single integrated data structure named *super-top-k list*, for representing all predicted top-k results of future windows. Objects are sorted based on their score in the super-top-k list, and each object has starting and ending window marks which show a set of windows in which the object participates in top-k result. To efficiently handle new arrival of objects, they define a *lower bound pointer (lbp)* for each window, which points to the object with the smallest score in the top-k list of the window. LBP set contains pointers for all the active windows.

---

[4]Note that the notion of candidate set in MTK is different from the one presented in [8].

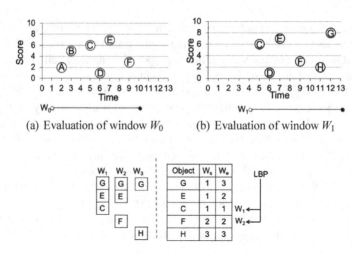

(a) Evaluation of window W_0　　　　(b) Evaluation of window W_1

(c) Independent predicted top-k result vs. integrated list at evaluation of window w_1

Fig. 2.2 The example that shows the objects in top-k result after join clause evaluation of windows W_0, and W_1

Figure 2.2 shows an example of data items in a stream and how Yang et al.'s approach evaluates the queries. Figure 2.2a, b shows a portion of a stream between time 0 and 13. The X axis shows the arriving time of each data item to the system, while the Y axis shows the score related to the data item. For the sake of clarity, each point is labeled in the Cartesian space with an ID. This stream is observed through a window that has a length equal to 9 units of time and slides every 3 units of time. In particular, Fig. 2.2a shows the content of window W_0 that opens at 1 and closes at 10. Figure 2.2b shows the next window W_1 after the sliding of 3 time units.

During window W_0 items A, B, C, D, E, and F come to the system (Fig. 2.2a). When W_0 expired, items A and B go out of the result. Before the end of window W_1, the new items G, and H appear (Fig. 2.2b). Evaluating query in Listing 1.1, give us item E as the top-1 result for window W_0 and item G as the result for window W_1.

Consider the above example, when asked to report the top-3 objects for each window. Figure 2.2c shows the content of the super-top-k list at the evaluation of window W_1. The left side of the picture shows the top-k result for each window. For instance, objects G, E, and C are in the top-3 result of window W_1 and objects G, E, and F are in the top-3 predicted result of window W_2. The right side shows the Super-top-k list which is a compact integrated list of all top-k results. Objects are sorted based on their scores. W_s, and W_e are starting- and ending-window-marks, respectively. The *lbp* of W_1, and W_2 are available, as those windows have top 3 objects in their predicted results.

The MinTopk algorithm consists of two maintenance steps: handling the expiration of the objects at the end of each window, and handling the insertion of new arrival objects.

For handling expiration, the top-k result of the expired window must be removed from the super-top-k list. The first k objects in the list with the highest score are the top-k result of the expired window. So, logically purging the first top-k objects of the super-top-k list is sufficient for handling expiration. It is implemented by increasing the starting-window-mark by 1, which means that the object will not be in the top-k list of the expired window any more. If the starting-window-mark becomes larger than the end-window-mark, the object will be removed from the list and the LBP set will be updated if any *lbp* points to the removed object.

For insertion of a new object, first the algorithm checks if the new object has the potential to become part of the current or the future top-k results. If all the *predicted top-k result* lists have k elements, and the score of the new object is smaller than any object in the super-top-k list, the new object will be discarded. If those lists have not reached the size of k yet, or if the score of the new object is larger than any object in the super-top-k list, the new object will be inserted in the super-top-k list based on its score. The starting and ending window marks will also be calculated for the new object. In the next step, for each window, in which the new object is inserted, the object with the lowest score, which is pointed by *lbp*, will be removed from the *predicted top-k result*. Like for the purging process, the starting-window-mark is increased by 1 and if it becomes larger than ending window mark, the object physically is removed from the super-top-k list and the LBP set will be updated if any *lbp* points to the removed object. In order to update *lbp* pointer, the algorithm simply moves it one position up in the super-top-k list.

The CPU complexity for MinTopK algorithm is $O(N_{new} * (log(MTK.size)))$ in the general case, with $O(N_{new})$ the number of new objects that come in each window, and $MTK.size$ is the size of the super-top-k list. The memory complexity in the general case is equal to $O(MTK.size)$. In the average case, the size of the super-top-k list is equal to $O(2k)$. So, in the average case, the CPU complexity is $O(N_{new} * (log(k)))$ and the memory complexity is $O(k)$. The authors also prove the optimality of the MinTopK algorithms. The experimental studies [38] on real streaming data confirm the out-performance of MinTopK algorithms over the previous solution [21].

2.6 Metrics

Measuring the accuracy of top elements [5] in the result is crucial in this book as well as for many applications such as information retrieval systems, search engines, and recommendation systems. In the last decades, different criteria have been introduced to measure this quality such as the precision at k, the accuracy at k, the normalized discounted cumulative gain ($nDCG$), or the mean reciprocal rank (MRR).

This section introduces various metrics that are used in this book in order to compare the possibly erroneous result of a query, named $Ans(Q_i)$, with certainly correct answers obtained from setting up an Oracle, named $Ans(Oracle)$.

Consider the following example which is used to explain the following metrics. For a given query (Q_i), the list of data items $\{A, B, C, D, E, F\}$ has the relevancy respectively equal to $\{6, 5, 4, 3, 2, 1\}$. So, the top-3 correct result of the query is equal to $Ans(Oracle) = \{A, B, C\}$. In addition, two top-3 erroneous answers of the query $(\{A, D, F\}$, and $\{F, C, B\})$ are considered as case 1, and case 2.

Jaccard Distance. Given that the result of the query that is used in the experiments of Chaps. 4, and 5, is a set of users' IDs, Jaccard distance is used to measure diversity of the set generated by the query and the one generated by the Oracle.

The *Jaccard index* is commonly used for comparing the similarity and diversity of overlapping sets (e.g., S_1 and S_2). The Jaccard index J is defined as the size of the intersection divided by the size of the union of the sets as follows:

$$J(S_1, S_2) = \frac{|S_1 \cap S_2|}{|S_1 \cup S_2|} \qquad (2.14)$$

The *Jaccard distance* d_J, which measures dissimilarity between sets, is complementary to the Jaccard index and is obtained by subtracting the Jaccard index from 1:

$$d_J(S_1, S_2) = 1 - J(S_1, S_2) \qquad (2.15)$$

Considering the previous example, where $Ans(Oracle) = \{A, B, C\}$, and $Ans(Q_i) = \{A, D, F\}$ (case 1), the Jaccard index of these two sets is computed as following:

$$J(Ans(Oracle), Ans(Q_i)) = \frac{|\{A\}|}{|\{A, B, C, D, F\}|} = \frac{1}{5} = 0.2$$

The Jaccard distance is computed as following:

$$d_J(Ans(Oracle), Ans(Q_i)) = 1 - J(Ans(Oracle), Ans(Q_i)) = 1 - 0.2 = 0.8$$

Respectively, for case 2 the Jaccard index is equal to 0.5, and the Jaccard distance is equal to 0.5. The result shows that case 1 has higher diversity with the correct answer, which means it has more errors.

Discounted Cumulative Gain. Discounted Cumulative Gain (DCG) is used widely in information retrieval to measure relevancy (i.e., the quality of ranking) for Web search engine algorithms. DCG applies a discount factor based on the position of the items in the list. DCG at particular position k is defined as:

$$DCG@k = \sum_{i=1}^{k} \frac{2^{rel_i} - 1}{\log_2(i + 1)} \qquad (2.16)$$

Considering the relevancy of each data items from the previous example the DCG of case 1 is computed as follows:

$$DCG@3 = \frac{63}{1} + \frac{7}{1.585} + \frac{1}{2} = 67.92$$

In order to compare different result sets for various queries and positions, DCG must be normalized across queries. First, the maximum possible DCG through position k is produced, which is called *Ideal DCG* ($IDCG$). This is done by sorting all relevant documents by their relative relevance. Then, the normalized discounted cumulative gain ($nDCG$), is computed as:

$$nDCG@k = \frac{DCG@k}{IDCG_{@k}} \tag{2.17}$$

In previous example, considering $\{A, B, C\}$ as the correct result, the $IDCG$ is computed as follows:

$$IDCG@3 = \frac{63}{1} + \frac{31}{1.585} + \frac{15}{2} = 90.06$$

and $nDCG@3$ for case 1 is computed as :

$$nDCG@3 = \frac{DCG}{IDCG} = \frac{63 + 4.42 + 0.5}{90.06} = 0.754$$

Precision. Precision in information retrieval indicates the ratio between the correct instances in the result and all the retrieved instances.

Considering the definition of precision as:

$$precision = \frac{tp}{tp + fp}, \tag{2.18}$$

where tp is the number of true positive values (the correct results), and fp is the number of false positive ones (the wrong results), the precision at position k is defined as:

$$precision@k = \frac{\# \text{ positive instances in the result}}{k} \tag{2.19}$$

where the true positive value is equal to the number of positive instances in the top-k result, and the summation of true positive and false positive values are equal to k.

The $precision@3$ for case 1 of the example is computed as :

$$precision@3 = \frac{tp}{tp + fp} = \frac{1}{3} = 0.333$$

So, for the first case, $nDCG@3$ is equal to 0.754 while $precision@3$ is equal to 0.33, which shows that the result is more relevant and less accurate. Data item A which is the most relevant item, is ranked in the accurate place, and the other answers are not the correct ones. On the contrary, the second case contains more correct answers, so a high value of $precision@3$ is expected. For the second case, $nDCG@3$ is equal to 0.288 while $precision@3$ is equal to 0.667, which indicates that the result is more accurate and less relevant. There are 2 correct answers in the result, but comparing to the case 1, they are less relevant.

References

1. Acosta M, Vidal M-E, Lampo T, Castillo J, Ruckhaus E (2011) Anapsid: an adaptive query processing engine for sparql endpoints. In: International semantic web conference. Springer, pp 18–34
2. Aranda CB, Arenas M, Corcho Ó, Polleres A (2013) Federating queries in SPARQL 1.1: syntax, semantics and evaluation. J Web Sem 18(1):1–17
3. Arasu A, Babu S, Widom J (2003) Cql: a language for continuous queries over streams and relations. In: International workshop on database programming languages. Springer, pp 1–19
4. Borzsony S, Kossmann D, Stocker K (2001) The skyline operator. In: Proceedings of the 17th international conference on data engineering. IEEE, pp 421–430
5. Boyd S, Cortes C, Mohri M, Radovanovic A (2012) Accuracy at the top. In: Advances in neural information processing systems, pp 953–961
6. Buil-Aranda C, Arenas M, Corcho O, Polleres A (2013) Federating queries in sparql 1.1: syntax, semantics and evaluation. Web Semant: Sci Serv Agents World Wide Web 18(1):1–17
7. Chang Y-C, Bergman L, Castelli V, Li C-S, Lo M-L, Smith JR (2000) The onion technique: indexing for linear optimization queries. In: ACM sigmod record, vol 29. ACM, pp 391–402
8. Dehghanzadeh S, Dell'Aglio D, Gao S, Della Valle E, Mileo A, Bernstein A (2015) Approximate continuous query answering over streams and dynamic linked data sets. In: 15th international conference on web engineering, Switzerland, Jun 2015
9. Dell'Aglio D, Della Valle E, Calbimonte J-P, Corcho Ó (2014) RSP-QL semantics: a unifying query model to explain heterogeneity of RDF stream processing systems. Int J Semant Web Inf Syst 10(4):17–44
10. Dell'Aglio D, Della Valle E, van Harmelen F, Bernstein A (2017) Stream reasoning: a survey and outlook. Data Sci (Preprint):1–25
11. Dobrzelecki B, Krause A, Hume AC, Grant A, Antonioletti M, Alemu TY, Atkinson M, Jackson M, Theocharopoulos E (2010) Integrating distributed data sources with ogsa–dai dqp and views. Philos Trans Royal Soc Lond A: Math Phys Eng Sci 368(1926):4133–4145
12. Fagin R (1999) Combining fuzzy information from multiple systems. J Comput Syst Sci 58(1):83–99
13. Fagin R, Lotem A, Naor M (2003) Optimal aggregation algorithms for middleware. J Comput Syst Sci 66(4):614–656
14. Ilyas IF, Aref WG, Elmagarmid AK (2002) Joining ranked inputs in practice. In: Proceedings of the 28th international conference on Very Large Data Bases. VLDB Endowment, pp 950–961
15. Ilyas IF, Aref WG, Elmagarmid AK (2004) Supporting top-k join queries in relational databases. The VLDB J Int J Very Large Data Bases 13(3):207–221
16. Ilyas IF, Beskales G, Soliman MA (2008) A survey of top-k query processing techniques in relational database systems. ACM Comput Surv (CSUR) 40(4):11
17. Ilyas IF, Shah R, Aref WG, Vitter JS, Elmagarmid AK (2004) Rank-aware query optimization. In: Proceedings of the 2004 ACM SIGMOD international conference on Management of data. ACM, pp 203–214

18. Jackson MJ, Antonioletti M, Dobrzelecki B, Hong NC (2011) Distributed data management with ogsa–dai. In: Grid and cloud database management. Springer, pp 63–86
19. Li C, Chen-Chuan Chang K, Ilyas IF, Song S (2005) Ranksql: query algebra and optimization for relational top-k queries. In: Proceedings of the 2005 ACM SIGMOD international conference on Management of data. ACM, pp 131–142
20. Magliacane S, Bozzon A, Della Valle E (2012) Efficient execution of top-k sparql queries. Semant Web–ISWC 2012, pp 344–360
21. Mouratidis K, Bakiras S, Papadias D (2006) Continuous monitoring of top-k queries over sliding windows. In: Proceedings of the 2006 ACM SIGMOD international conference on Management of data. ACM, pp 635–646
22. Natsev A, Chang Y-C, Smith JR, Li C-S, Vitter JS (2001) Supporting incremental join queries on ranked inputs. In: VLDB, vol 1, pp 281–290
23. Papadias D, Tao Y, Greg F, Seeger B (2005) Progressive skyline computation in database systems. ACM Trans Database Syst (TODS) 30(1):41–82
24. Pérez J, Arenas M, Gutierrez C (2009) Semantics and complexity of SPARQL. ACM Trans Database Syst 34(3)
25. Pérez J, Arenas M, Gutierrez C (2009) Semantics and complexity of sparql. ACM Trans Database Syst (TODS) 34(3):16
26. Pripužić K, Žarko IP, Aberer K (2015) Time-and space-efficient sliding window top-k query processing. ACM Trans Database Syst (TODS) 40(1):1
27. Prud'hommeaux E, Seaborne A (2008) SPARQL Query Language for RDF. W3C Recommendation, January 2008. http://www.w3.org/TR/rdf-sparql-query/
28. Quilitz B, Leser U (2008) Querying distributed rdf data sources with sparql. In: European semantic web conference. Springer, pp 524–538
29. Schnaitter K, Polyzotis N (2008) Evaluating rank joins with optimal cost. In: Proceedings of the twenty-seventh ACM SIGMOD-SIGACT-SIGART symposium on Principles of database systems. ACM, pp 43–52
30. Schwarte A, Haase P, Hose K, Schenkel R, Schmidt M (2011) Fedx: optimization techniques for federated query processing on linked data. In: International semantic web conference. Springer, pp 601–616
31. Urhan T, Franklin MJ (2000) Xjoin: a reactively-scheduled pipelined join operatorỳ. Bull Techn Comm pp 27
32. Wagner A, Bicer V, Tran T (2014) Pay-as-you-go approximate join top-k processing for the web of data. In: European semantic web conference. Springer, pp 130–145
33. Wagner A, Duc TT, Ladwig G, Harth A, Studer R (2012) Top-k linked data query processing. In: Extended semantic web conference. Springer, pp 56–71
34. Wang D, Zou L, Zhao D (2015) Top-k queries on rdf graphs. Inf Sci 316:201–217
35. Wiederhold G (1992) Mediators in the architecture of future information systems. Computer 25(3):38–49
36. Wiederhold G (1992) Mediators in the architecture of future information systems. Computer 25(3):38–49
37. Wood D, Lanthaler M, Cyganiak R (2014) RDF 1.1 concepts and abstract syntax, February 2014
38. Yang D, Shastri A, Rundensteiner EA, Ward MO (2011) An optimal strategy for monitoring top-k queries in streaming windows. In: Proceedings of the 14th international conference on extending database technology. ACM, pp 57–68
39. Yang S, Han F, Wu Y, Yan X (2016) Fast top-k search in knowledge graphs. In: Data engineering (ICDE), 2016 IEEE 32nd international conference on. IEEE, pp 990–1001
40. Yi K, Yu H, Yang J, Xia G, Chen Y (2003) Efficient maintenance of materialized top-k views. pp 189–200
41. Zhang Z, Hwang S-W, Chen-Chuan Chang K, Wang M, Lang CA, Chang Y-C (2006) Boolean+ ranking: querying a database by k-constrained optimization. In: Proceedings of the 2006 ACM SIGMOD international conference on Management of data. ACM, pp 359–370

Chapter 3
ACQUA: Approximate Continuous Query Answering in RSP

Abstract This chapter focuses on continuous query answering over an RDF stream and evolving distributed data in the context of RDF Stream Processing (RSP) applications. Although RSP engines are suitable for continuous query evaluation, complex analyses that combine data streams and distributed data is challenging. This chapter introduces an approach in which a replica of the distributed data is stored locally. In addition, different maintenance processes are proposed to keep the data in the replica fresh, and to minimize (or even avoid) the errors in the query result. This explains the name ACQUA given to the framework. It offers a possibly approximate solution to the problem of Continues QUery Answering. Specifically, this chapter presents ACQUA architecture and policies tailored for basic graph patterns.

Keywords Continuous query answering · Streaming data · Evolving distributed data · RDF data · RSP engine

3.1 Introduction

RDF Stream Processing (RSP) applications are becoming popular in different domains such as social networks, and Web of Things. Although RSP engines can deal with the variety and velocity of data streams, complex analyses with combining data streams and evolving distributed data are challenging. In a Semantic Web setting, distributed data is usually exposed through SPARQL endpoints [3].

Current RSP languages, like C-SPARQL [2], SPARQL$_{stream}$ [4], CQELS-QL [12] support queries involving streaming and distributed data, but, implementations of those languages (RSP engines) need invoking the remote services for each query evaluation, without any optimization. So, the query evaluation may generate high loads on remote services and decrease the response times. The time to access and fetch the distributed data can be so high to put the RSP engine at risk of violating the reactiveness requirement in continuous query answering. Therefore, optimization techniques are highly needed to provide faster responses to this class of continuous queries.

© The Author(s), under exclusive license to Springer Nature Switzerland AG 2020   27
S. Zahmatkesh and E. Della Valle, *Relevant Query Answering over Streaming and Distributed Data*, PoliMI SpringerBriefs,
https://doi.org/10.1007/978-3-030-38339-8_3

```
1   REGISTER STREAM <:Influencers>
2   AS CONSTRUCT {?user a :influentialUser}
3   WHERE {
4       WINDOW :W(10m,1m) ON :S {?user :hasMentions ?mentionsNumber}
5       SERVICE :BKG {?user :hasFollowers ?followersCount }
6   }
```

Listing 3.1 Sketch of the query with 1:1 join relationship

Consider Example 1 introduced in Chap. 1: a Web advertising company; that wants to continuously detect influential Social Network users in order to ask them to endorse its commercials. Listing 3.1 shows how the example can be declared as a continuous RSP-QL query using the syntax proposed in [6]. Lines 1, and 2 register the query, and describe how to construct the results. Line 4, every minute, selects from a window opened on the stream :S the users mentioned in the last 10 min. Line 5 asks the remote service :BKG to select the number of followers for the users mentioned in the window.

A possible solution is to store the intermediate results of SERVICE clauses in the local replica inside the engine instead of pulling data from remote SPARQL endpoints at each evaluation. However, when time passes, the distributed data in the remote service changes and updates are not reflected in the local replica, so, the freshness of the data and the accuracy of the answer decrease. To overcome this problem, a maintenance process is introduced, which identifies the stale data items in the local replica and replaces them with fresh values from the remote services.

This chapter reports on the investigation of the following research question:

RQ.1 *Given a query that joins stream data with distributed data, how the local replica of distributed data can be refreshed in order to guarantee reactiveness while maximizing the accuracy of the continuous answer?*

The proposed solution is a maintenance process based on the following consideration: the accuracy of the current response is not affected by refreshing elements that are fresh or not involved in the answer of the current query. Thus, an efficient maintenance process should refresh local replica entries that are both stale and involved in the current query evaluation. The research question is investigated through two hypotheses.

Hp.1.1   The accuracy of the answer can increase by maintaining part of the local replica involved in the current query evaluation.

Hp.1.2   The accuracy of the answer increases by refreshing the (possibly) stale entries in local replica that would remain fresh in the highest number of evaluations.

Elements of local replica that are involved in the current evaluation depend on the content of the stream in the current evaluation. In the first step, a join method named *Window Service Join* (WSJ) is proposed to filter out local replica elements that are not involved in the current evaluation, as their maintenance does not affect the accuracy of the response.

In the next step, *Window Based Maintenance* (WBM) *policy* is proposed, which is a policy that assigns a score to the local replica elements based on the estimated best before time, i.e., the time on which a fresh element is estimated to become stale, and the number of next evaluations that the item is going to be involved in. The former is possible by exploiting the change frequency of elements, while the second exploits the streaming part of the query and the window operator to (partially) foresee a part of the future answers.

The remainder of the chapter is organized as follows. Section 3.2 formalizes the problem. Section 3.3 introduces the proposed solution for refreshing the local replica of distributed data and discusses the maintenance policy in detail. Section 3.4 reports on experimental evaluation of the proposed approach, and finally, Sect. 3.6 concludes.

3.2 Problem Statement

This chapter considers continuous join queries over a data stream and a distributed dataset, and here is the intuition: the RSP engine must avoid accessing the whole distributed data at each evaluation. Instead, it uses a local replica of the distributed data and keeps it fresh using a maintenance process that refreshes only a minimum subset of the replica.

In order to guarantee the reactiveness of the system, the maximum number of fetches (namely a **refresh budget** denoted with γ) should be given to the RSP engine. If γ fetches are enough to refresh all stale data of replica, the RSP engine gives the correct answer, otherwise, some data becomes stale and it gives an approximated answer.

The problem introduced in the previous section can be modeled in the context of an RSP engine E. Given a query Q, the result of continuously evaluating query Q over time denotes by $Ans(Q)$, and the expected result denotes by $Ans(Oracle)$. The accuracy of the result $(acc(E, Q))$ is defined as a ratio between the number of elements of $Ans(Q)$ that are also in $Ans(Oracle)$, and the total number of elements in $Ans(Oracle)$. The query latency is the time required to process the query answer. For RSP engines, an adaptation is needed and the query latency is a set of values, each related to an evaluation. $lat(E, Q)$ denotes the latency of the current evaluation of query Q.

A continuous query Q executes over an RDF stream S and a remote SPARQL endpoint BKG with some QoS constraints (α, ρ), i.e., the answer should have an accuracy equals or greater than α and should be provided at most in ρ time units. The output of the evaluation $Ans(Q)$ is the sequence of answers produced by the engine continuously evaluating query Q over time. The QoS constraints can be expressed in the following way:

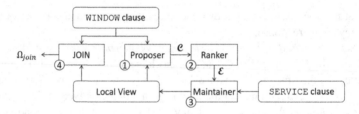

Fig. 3.1 The ACQUA framework proposed in [5] to address the problem of joining streaming and distributed data

$$(acc(E, q) > \alpha) \wedge (lat(E, q) < \rho) \text{for each evaluation} \qquad (3.1)$$

Specifically, a class of RSP-QL [7] queries is considered where the streaming data is obtained by a window identified by the IRI u_S, the distributed data is available via a SPARQL service at the URL u_B and the algebraic expression SE contains the following graph patterns:

$$(WINDOW \; u_S \; P_S) \; JOIN \; (SERVICE \; u_B \; P_B) \qquad (3.2)$$

The goal is to design a maintenance process that minimizes the number of stale elements involved in the computation of the current answer, and so decreases the errors in the response. The cost of the maintenance process w.r.t. the constraints on responsiveness and accuracy, also should be minimized.

3.3   Proposed Solution

This section introduces the proposed solution for Approximate Continuous Query Answering (shortly name ACQUA in the reminder of book). ACQUA introduces a replica $\mathcal{R}$ to store the result of $(SERVICE \; u_B \; P_B)$ as a set of solution mapping. As stated in Chap. 2, the evaluation of SPARQL query is represented as a set of solution mapping. A *maintenance process* is introduced to keep $\mathcal{R}$ up-to-date. Figure 3.1 depicts the elements that compose the maintenance process: a proposer, a ranker and a maintainer. (1) The *proposer* selects a set $\mathcal{C}$ of candidate mappings for the maintenance from local replica $\mathcal{R}$. As the accuracy of the answer depends on the freshness of the mapping that involves in the current window, the maintenance process should focus of them. (2) The *ranker* orders $\mathcal{C}$ by using some relevancy criteria. The algorithm that provides the top γ elements for the next step named *maintenance policy*. (3) The *maintainer* refreshes the top γ elements of $\mathcal{C}$ (the elected set $\mathcal{E}$), where γ is named *refresh budget* and encodes the number of requests the RSP engine can submit to the remote services without losing reactiveness. After the maintenance, (4) the join operation is performed.

Algorithm 3.1: The WSJ next method

---

1 **if** *first iteration* **then**
2   **while** *WinOp has next* **do**
3 | append WinOp.next() to Ω_{window} ;
4   **end**
5      $\mathcal{C}$=$\mathcal{R}$.compatibleMapping(Ω_{window}) ;
6      $M(\mathcal{R}, \gamma)$;
7      $it = \Omega_{window}$.iterator() ;
8 **end**
9 **if** *it is not empty* **then**
10   μ^W = it.next() ;
11      $\mu^{\mathcal{R}}$=$\mathcal{R}$.compatibleMapping(μ^W) ;
12   return $\mu^W \cup \mu^{\mathcal{R}}$;
13 **end**

The solution is implemented in a system composed of the Window Service Join method (WSJ) and the Window Based Maintenance policy (WBM). WSJ performs the join and starts the maintenance process as a proposer, while WBM completes the maintenance process by ranking the candidate set (as ranker) and maintaining the local replica (as maintainer).

3.3.1 Window Service Join Method

WSJ method starts the maintenance process as proposer by performing the join between streaming and distributed data. In designing WSJ, in order to be responsive, the latency is fixed based on constraint ρ, and the accuracy of the answer is maximized accordingly. To deal with the responsiveness requirement, the notion of refresh budget γ is introduced as the number of elements in $\mathcal{R}$ that can be maintained at each evaluation. WSJ builds the candidate set by selecting the mappings in $\mathcal{R}$ compatible with the ones from the evaluation of $(WINDOW \ u_S \ P_S)$.

Algorithm 3.1 shows pseudo-code of the WSJ method. The first invocation of next function retrieves the results of Ω_{join}. The WINDOW expression is evaluated and the bag of solution mappings Ω_{window} is retrieved from the WinOp operator (Lines 2–4). WSJ computes the candidate set $\mathcal{C}$ as the set of mappings in $\mathcal{R}$ compatible with the ones in Ω_{window} (Line 5). The maintenance policy M gets set $\mathcal{C}$ and the refresh budget γ as inputs (Line 6) in order to refresh the local replica. Then, an iterator is initiated over Ω_{window} (Line 7). At the end, the join is performed (Lines 9–13) between each mapping in Ω_{window} and the compatible mapping from $\mathcal{R}$ and the result is returned at each invocation of next function.

3.3.2   Window Based Maintenance Policy

WBM policy elects the mappings to be refreshed and maintains the local replica. Its goal is to maximize the accuracy of the query answer, given that it can refresh at most γ mappings of the local replica at each evaluation.

To determine if a mapping in $\mathcal{C}$ is fresh or stale, an access to the remote SPARQL endpoint BKG is required, and it is not possible (as explained above). So, WSJ-WBM policy uses the *best before time* to decide which mapping would affect the freshness of query results the most and the longest. That means, WBM orders the candidate set assigning to each mapping $\mu_i \in \mathcal{C}$ a score defined as:

$$score_i(t) = min(L_i(t), V_i(t)), \tag{3.3}$$

where t is the evaluation time, $L_i(t)$ is the *remaining life time*, i.e. the number of future evaluations that involve the mapping, and $V_i(t)$ is the *normalized renewed best before time*, i.e., the renewed best before time normalized with the sliding window parameters. The intuition behind WBM is to prioritize the refreshing of the mappings that contribute the most to the freshness in the current and next evaluations. That means, WBM identifies the mappings that are going to be used in the upcoming evaluations and that allows saving future refresh operations.

Given a sliding window $\mathbb{W}(\omega, \beta)$, L_i and V_i are defined as:

$$L_i(t) = \left\lceil \frac{t_i + \omega - t}{\beta} \right\rceil, \tag{3.4}$$

$$V_i(t) = \left\lceil \frac{\tau_i + I_i(t) - t}{\beta} \right\rceil \tag{3.5}$$

where t_i is the time instant associated with the mapping μ_i, τ_i is the current best before time, and $I_i(t)$ is the change interval, that captures the remaining time before the next expiration of μ_i. It is worth noting that I_i is potentially unknown and could require an estimator.

Algorithm 3.2 shows the pseudo-code of this maintenance policy. First, WBM identifies the possibly stale mappings. Then, it assigns a score to the possibly stale elements $\mathcal{PS}$ (Lines 2–6) to create priority for limited refresh budget. The score is used to order the mappings (Line 7). WBM builds the set of elected mappings $\mathcal{E} \subset \mathcal{PS}$ to be refreshed, by getting the top *gamma* ones (Lines 7–8). Finally, the refresh is applied to maintain local replica (Lines 8–13): for each mapping of $\mathcal{E}$, WBM invokes the SERVICE operator to retrieve from the remote SPARQL endpoint the fresh mapping and replaces it in the local replica. Additionally, the best before time values of the refreshed elements are also updated.

Figure 3.2 shows an example that illustrates how WSJ-WBM policy works. Figure 3.2a shows the mappings that enter the window clause between time 1 and 12. Each window has a length of 5 units of time and slides every 2 units of time. For instance, window W_0 opens at 1 and closes at 6 (excluded). Each map-

---

Algorithm 3.2: The pseudo-code of M method

---

1   $\mathcal{PS}$= possible stale elements of $\mathcal{C}$;

2 **foreach** $\mu^{\mathcal{R}} \in \mathcal{C}$ **do**

3 | compute the remaining life time of $\mu^{\mathcal{R}}$;

4 | compute the renewed best before time of $\mu^{\mathcal{R}}$;

5 | compute the score of $\mu^{\mathcal{R}}$;

6 **end**

7 order $\mathcal{PS}$ w.r.t. the scores;

8 $\mathcal{E}$ = first γ mappings of $\mathcal{PS}$;

9 **foreach** $\mu^{\mathcal{R}} \in \mathcal{E}$ **do**

10 | μ^{S} = ServiceOp.next(JoinVars($\mu^{\mathcal{R}}$));

11 | replace $\mu^{\mathcal{R}}$ with μ^{S} in $\mathcal{R}$;

12 | update the best before time of $\mu^{\mathcal{R}}$;

13 **end**

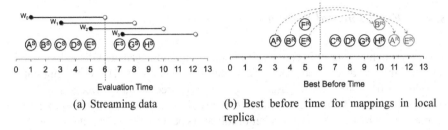

(a) Streaming data   (b) Best before time for mappings in local replica

Fig. 3.2 The example that shows how WSJ-WBM policy works

ping is marked with a point and for the sake of clarity, each point is labeled with I^{S} where I is the ID of the subject of mapping and S indicates that the mappings appear on the data stream. So, for example during window W_0 mappings A^{S}, B^{S}, C^{S}, D^{S}, and E^{S} appear on the data stream.

Figure 3.2b shows the mappings in the local replica. The mappings in the replica are indicated by R. The replica contains mappings A^{R}, B^{R},..., H^{R}. The x-axis shows the value of *best before time* for each mapping. It is worth to note that points with the same ID in Fig. 3.2a, and b indicate compatible mappings.

At the end of window W_0, at time 6, WSJ computes the candidate set by selecting compatible mappings with those in the window. The candidate set $\mathcal{C}$ contains mappings A^{R}, B^{R}, C^{R}, D^{R}, and E^{R}. In the next step, WBM finds the possible stale mappings by comparing their best before time values with the current time. The possibly stale mappings are $\mathcal{PS} = \{A^{R}, B^{R}, E^{R}\}$. The best before time of other mappings are greater than the current time, so they do not need to be refreshed.

The remaining life time shows the number of successive evaluations for each mapping. The remaining life time of mapping A^{R}, B^{R}, E^{R} are 1, 1 and 3, respectively. Figure 3.2b shows the renewed best before time of the elements in $\mathcal{PS}$ by the arrows. The normalized renewed best before time ($V_i(t)$) of mappings A^{R}, B^{R}, and E^{R}

at time 6 are respectively 3, 2 and 3. Finally, **WBM** computes the score for each mapping at time 6: $score_A(6) = 1$, $score_B(6) = 1$, and $score_E(6) = 3$. Given the refresh budget γ equal to 1, the elected mapping will be E^R, which has the highest score.

3.4 Experiments

This section reports the result of experiments that study the performance of **WSJ** and **WBM** to validate the hypotheses presented in Sect. 3.1. Section 3.4.1 introduces the experimental setting that is used to check the validity of the hypotheses. Sections 3.4.2, and 3.4.3 investigate hypotheses Hp.1.1, and Hp.1.2 respectively.

3.4.1 Experimental Setting

The experimental dataset is composed of streaming and distributed data. Two data sets are built for distributed data: a real, and a synthetic one. The streaming data is collected from 400 verified users of Twitter for three hours of tweets using the streaming API of Twitter. The real distributed data is collected by invoking the Twitter API, which returns the number of followers per user, every minute during the three hours of recording of the streaming data. The snapshots of the users' follower numbers keep track of the changes in order to replay the evolution of the distributed data. Additionally, the synthetic distributed data is built by assigning a different change rate to each user and changing the follower number accordingly. The test query performs the join between the collected stream and distributed data. The query uses a window that slides every 60 seconds.

3.4.2 Experiment 1—Comparison of Proposers

This experiment investigates the first hypothesis by comparing **WSJ** as join method with a proposer in which the whole replica is used as the candidate set, i.e., it does not consider the query, and the compatible mappings in the maintenance process. This proposer is named General proposer (**GNR**).

In order to complete the maintenance process, two maintenance policies are proposed, which are inspired by the random (**RND**) and Least-Recently Used (**LRU**) cache replacement algorithms [10]. RND policy randomly ranks the mappings in the candidate set; LRU policy orders the candidate set $\mathcal{C}$ by the time of the last refresh of the mappings: the more recently a mapping has been used in a query, the higher is its rank.

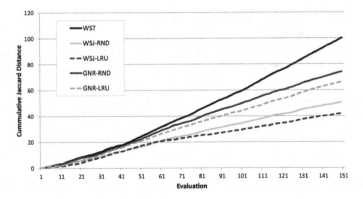

Fig. 3.3 Result of experiment that investigates Hypothesis Hp.1.1

So, combining these two maintenance policies with the two proposers results in four maintenance processes: GNR-LRU, GNR-RND, WSJ-LRU, and WSJ-RND. The study reported hereafter compares them. Moreover, it considers the worst maintenance process (WST) as a upper bound policy which does not refresh the local replica throughout the evaluations.

In order to investigate the hypothesis, an Oracle is organized that, at each iteration i, certainly provides correct answers $Ans(Oracle_i)$ and its answers can be compared with the possibly erroneous ones of the query $Ans(Q_i)$. Jaccard distance is used to measure diversity of the set generated by the query and the one generated by the Oracle, and cumulative Jaccard distance at the kth iteration $d_J^C(k)$ is introduced as:

$$d_J^C(k) = \sum_{i=1}^{k} d_J(Ans(Q_i), Ans(Oracle_i)) \qquad (3.6)$$

where $d_J(Ans(Q_i), Ans(Oracle_i))$ is the Jaccard distance of iteration i.

Figure 3.3 shows the results of the experiment, which compares WSJ with GNR. Two maintenance policies LRU, and RND are considered to complete the maintenance process. In this experiment, the refresh budget γ is equal to 3, and it runs 140 iterations of the query evaluation. The chart shows the cumulative Jaccard distance over 140 evaluations (the lower, the better performance). WST is the upper bound policy, which does not have any proposer, and does not update the replica, so, it has the highest error. The figure also shows that WSJ performs significantly better than GNR in both cases of combining with LRU, and RND policies.

In order to check if the result generalizes, the experiment is repeated with different refresh budgets. First, the average dimension of the candidate sets is computed, then the refresh budget is set equal to 8, 15 and 30% of the average value (respectively 3, 5 and 10).

Table 3.1 reports the result of experiments for different values of refresh budget on the synthetic and the real dataset by average freshness. The result shows that

Table 3.1 Freshness comparison of WSJ in synthetic/real datasets

γ	Synthetic dateset					Real dataset				
	WST	GNR RND	GNR LRU	WSJ RND	WSJ LRU	WST	GNR RND	GNR LRU	WSJ RND	WSJ LRU
3	0.23	0.26	0.27	0.4	0.38	0.30	0.34	0.33	0.46	0.47
5	0.23	0.26	0.28	0.48	0.51	0.30	0.36	0.35	0.57	0.58
10	0.23	0.32	0.33	0.64	0.76	0.30	0.41	0.41	0.68	0.80

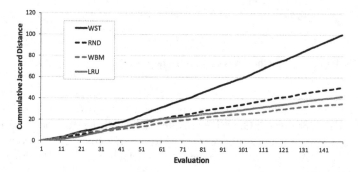

Fig. 3.4 Result of experiment that investigates Hypothesis Hp.1.2

when the refresh budget increases, WSJ has better improvements comparing to GNR. For instance, in the synthetic dataset GNR-RND improves from 0.26 to 0.32, while WSJ-RND improves from 0.4 to 0.64. It happens because WSJ chooses the mappings from those currently involved in the evaluation, while GNR chooses from the whole local replica.

3.4.3   Experiment 2—Comparison of Maintenance Polices

This section reports the result of experiments related to the second hypothesis, which evaluate the performance of WBM policy. This experiment compares WBM with other maintenance policies. WST policy is used as a upper bound maintenance process for the error. RND and LRU, which are presented in the previous section, are used as policies to make the comparison. For the completeness of the maintenance process, WSJ is used as a proposer because of its good performance.

In this experiment, the refresh budget is set equal to 3, and it runs 140 iterations of query evaluation. Figure 3.4 shows the cumulative Jaccard distance over multiple query evaluations of the experiment. As the result shows, WSJ-WBM policy outperforms other policies (WSJ-RND, and WSJ-LRU). As expected, the WST policy is always the worst one.

Table 3.2 Freshness comparison of WBM,RND, and LRU in synthetic/real datasets

γ	Synthetic dateset				Real dataset			
	WST	WSJ RND	WSJ LRU	WSJ WBM	WSJ	WST RND	WSJ LRU	WSJ WBM
3	0.23	0.39	0.38	0.46	0.30	0.45	0.47	0.52
5	0.23	0.49	0.50	0.60	0.30	0.57	0.58	0.61
10	0.23	0.64	0.76	0.81	0.30	0.68	0.80	0.80

The experiment is repeated with different refresh budgets, and the results are shown in Table 3.2. In general, **WSJ-WBM** shows better performance comparing to **WSJ-LRU**, and **WSJ-RND**. However, **WSJ-WBM** leads to higher improvements when the refresh budget is lower.

3.5 Related Work

This section reviews the related work in data source replication. Data sources replication is used by many systems to increase availability and reactiveness, however, maintenance processes are needed to keep the freshness of data and reduce inconsistencies. To get accurate answers and reduce inconsistencies, a maintenance process is needed to keep the local replicas fresh. Extensive studies exist about optimization and maintenance process in database community such as [1, 9, 11, 13, 14]. However, those works still do not consider the problem of combing streaming data with distributed datasets.

Babu et al. [1] address the problem of using caches to improve the performance of continuous queries. Authors propose an adaptive approach to handle changes of update streams, such as stream rates, data characteristics, and memory availability over time. The approach manages the trade-off between space and query response time. They propose an Adaptive Caching algorithm that estimates cache benefit and cost online in order to select and allocate memory to caches dynamically.

Guo et al. [9] develop a data quality-aware, finer grained cache model. They formally introduce fundamental cache properties: presence, consistency, completeness, and currency. In the proposed cache model, users can specify a cache schema by defining a set of local views, and their cache constraints to guarantee cache properties. Authors integrate consistency and currency checking into query optimization and evaluation. The optimizer checks most of the consistency constraints. Dynamic plan of query includes currency checks and inexpensive checks for dynamic consistency constraints that cannot be validated during optimization.

Labrinidis et al. [11] explore the idea that a trade-off exists between the quality of answers and time for the maintenance process. In the context of the Web, view materialization is an attractive solution, since it decouples the serving of access requests from the handling of the updates. They introduce the Online View Selection Problem and propose a way to dynamically select materialization views to maximize performance while keeping data freshness at a reasonable level. They propose an adaptive algorithm for Online View Selection Problem that decides to materialize or just cache views. Their approach is based on user-specified data freshness requirements.

Viglas et al. [14] propose an optimization in join query evaluation for inputs arriving in a streaming fashion. It introduces a multi-way symmetric join operator, in which inputs can be used to generate results in a single step, instead of a pipeline execution.

Umbrich et al. [13] address the response time and freshness trade-off in the Semantic Web domain. Cached Linked Data suffers from missing data as it covers partial of the resources on the Web, on the other hand, live querying has slow query response time. They propose a hybrid query approach that improves in both directions, by considering a broader range of resources than cashes, and offering faster result than live querying.

3.6   Conclusion

This chapter introduces ACQUA which approximately answers queries over data stream and distributed Web data. Reactiveness is the most important performance indicator for evaluating queries in RSP engines. When the distributed data evolves, the time to access it may exceed the time between two consecutive evaluations. So the RSP engine loses its reactiveness.

In order to guarantee reactiveness, an extended architecture of RSP engines is proposed with the following components: (*i*) a replica to store the distributed data at query registration time, (*ii*) a maintenance policy to keep the replica fresh, and (*iii*) a refresh budget to limit the number of access to the distributed data to guarantee reactiveness by design. In this way, accurate answers can be provided while meeting operational deadlines.

ACQUA policies apply to queries that join a basic graph pattern in a WINDOW clause with another basic graph pattern in a SERVICE clause, but it can be extended for different classes of queries.

Gao et al. [8] study the maintenance process for a class of queries that extends the 1:1 join relationship of ACQUA to M:N join. It models the join between streams and distributed data as a bipartite graph. The proposed algorithm employs the bipartite graph to model the join selectivity between streaming and distributed data, and updates data items with higher selectivity. Authors introduce two extensions of the maintenance process and investigate the best time for updating distributed data and propose flexible budget allocation method.

Although it is possible to use FILTER clauses in ACQUA framework, the proposed policies do not consider it in selecting the mappings to update. So, the policies may waist the refresh budget for updating mappings that do not satisfy the filtering condition, and cannot appear in the result set. The following chapter investigates this class of queries.

References

1. Babu S, Munagalat K, Widom J, Motwani R (2005) Adaptive caching for continuous queries. In: 2005 Proceedings of 21st International Conference on Data Engineering, ICDE 2005, pp 118–129. IEEE
2. Barbieri DF, Braga D, Ceri S, Valle ED, Grossniklaus M (2010) Querying rdf streams with c-sparql. ACM SIGMOD Record 39(1):20–26
3. Buil-Aranda C, Arenas M, Corcho O, Polleres A (2013) Federating queries in sparql 1.1: syntax, semantics and evaluation. Web Semant Sci Serv Agents World Wide Web 18(1):1–17
4. Calbimonte JP, Jeung H, Corcho O, Aberer K (2012) Enabling query technologies for the semantic sensor web. Int J Semant Web Inf Syst 8:43–63
5. Dehghanzadeh S, Dell'Aglio D, Gao S, Della Valle E, Mileo A, Bernstein A (2015) Approximate continuous query answering over streams and dynamic linked data sets. In: 15th International conference on web engineering, Switzerland, June 2015
6. Dell'Aglio D, Calbimonte JP, Della Valle E, Corcho O (2015) Towards a unified language for rdf stream query processing. In: International semantic web conference. Springer, Cham, pp 353–363
7. Dell'Aglio D, Della Valle E, Calbimonte JP, Corcho O (2014) RSP-QL semantics: a unifying query model to explain heterogeneity of RDF stream processing systems. Int J Semant Web Inf Syst 10(4):17–44
8. Gao S, Dell'Aglio D, Dehghanzadeh S, Bernstein A, Della Valle E, Mileo A (2016) Planning ahead: stream-driven linked-data access under update-budget constraints. In: International semantic web conference. Springer, Cham, pp 252–270
9. Guo H, Larson P-Å, Ramakrishnan R (2005) Caching with good enough currency, consistency, and completeness. In: 2005 Proceedings of the 31st international conference on Very large data bases. VLDB Endowment, pp 457–468
10. Koskela T, Heikkonen J, Kaski K (2003) Web cache optimization with nonlinear model using object features. Comput Netw 43(6):805–817
11. Labrinidis A, Roussopoulos N (2004) Exploring the tradeoff between performance and data freshness in database-driven web servers. VLDB J 13(3):240–255
12. Le-Phuoc D, Dao-Tran M, Parreira JX, Hauswirth M (2011) A native and adaptive approach for unified processing of linked streams and linked data. In: The semantic web–ISWC 2011. Springer, Berlin, pp 370–388
13. Umbrich J, Karnstedt M, Hogan A, Parreira JX (2012) Freshening up while staying fast: towards hybrid SPARQL queries. In: Knowledge engineering and knowledge management. Springer, Berlin, pp 164–174
14. Viglas SD, Naughton JF, Burger J (2003) Maximizing the output rate of multi-way join queries over streaming information sources. In: Proceedings of the 29th international conference on Very large data bases-Volume 29. VLDB Endowment, pp 285–296

Chapter 4
Handling Queries with a FILTER Clause

Abstract This chapter extends the ACQUA proposal presented in Chap. 3. It presents ACQUA.F, a new maintenance policy for an extended class of queries: the one that joins streaming data with the distributed data, and contains a filter constrain on distributed data. In addition, it explains how to combine this policy with the one proposed in the previous chapter. Finally, a set of experimental evaluations shows how the proposed policies guarantee reactiveness while keeping the replica fresh.

Keywords Continuous query answering · Streaming data · Evolving distributed data · RDF data · RSP engine · Filter constrain

4.1 Introduction

As stated in Chap. 1 the variety and the velocity of the Web of data is growing, and many Web applications require to continuously answer queries that combine dynamic data streams with data distributed over the Web. Consider the Example 1 introduced in Chap. 1: a Web advertising company; that wants to continuously detect influential Social Network users in order to ask them to endorse its commercials. It can be extended and encode in a continuous query like:

Example 3 Every minute give me the IDs of the users who are mentioned on Social Network in the last 10 min whose number of followers is greater than 100,000.

What makes continuously answering this query challenging is the fact that the number of followers of a user (in the distributed data) tends to change when she is mentioned (in the social stream), i.e., the value of the number of followers becomes stale faster. There may be users, whose number of followers was slightly below 100,000 in the last evaluation (and, thus, were not included in the last answer), who may now have slightly more than 100,000 followers (and, thus, are in the current answer).

If the application requires an answer every minute and fetching the current number of followers for a user (mentioned in the social stream) requires around 100

© The Author(s), under exclusive license to Springer Nature Switzerland AG 2020   41
S. Zahmatkesh and E. Della Valle, *Relevant Query Answering over Streaming and Distributed Data*, PoliMI SpringerBriefs,
https://doi.org/10.1007/978-3-030-38339-8_4

milliseconds,[1] just fetching this information for 600 users takes the entire available time. In other words, fetching all the distributed data may put the application at risk of losing reactiveness, i.e., it may not be able to generate an answer while meeting operational deadlines.

RSP engines presented in Chap. 2, offer a good starting point to satisfy this information need. The query above can be written as a continuous query for existing RSP engines. This query has to use a SERVICE clause[2] which is supported by C-SPARQL [1], SPARQL$_{stream}$ [2], CQELS-QL [5], and RSP-QL [4].

Listing 4.1 shows how the above example can be declared as a continuous RSP-QL query using the syntax proposed in [3]. Line 1 registers the query in the RSP engine. Line 2 describes how to construct the results at each evaluation. Line 4, every minute, selects from a window opened on the stream : S the users mentioned in the last 10 min. Line 5 asks the remote service : BKG to select the number of followers for the users mentioned in the window. Line 6 filters out, from the results of the previous join, all those users whose number of followers is above the 100,000 (namely, the Filtering Threshold).

```
1   REGISTER STREAM <:Influencers> AS
2   CONSTRUCT {?user a :influentialUser}
3   WHERE {
4       WINDOW :W(10m,1m) ON :S {?user :hasMentions ?mentionsNumber}
5       SERVICE :BKG {?user :hasFollowers ?followersCount }
6       FILTER (?followersCount > 100000)
7   }
```

Listing 4.1 Sketch of the query studied in the problem

As mentioned in Chap. 3, ACQUA policies were empirically demonstrated to be effective, but the approach focuses only on the JOIN and does not optimize the FILTER clause (at line 6), so it does not consider relevancy in the query. ACQUA policies may decide to refresh a mapping that will be discarded by the FILTER clause. In this case, ACQUA policies are throwing away a unit of budget.

This chapter, instead, investigates maintenance policies that explicitly consider the FILTER clause and exploit the presence of a Filtering Threshold that selects a subset of the mappings returned by the SERVICE clause. By avoiding to use the refresh budget to update mappings that will be discarded by the FILTER clause, the policies presented in this chapter have the potential to address the limits of ACQUA policies. This chapter answers the following research question:

RQ.2   *Given a query that joins streaming data returned from a WINDOW clause with filtered distributed data returned from a SERVICE clause how the local replica can be refreshed in order to guarantee reactiveness while maximizing the freshness of the mappings in the replica?*

---

[1] 100 ms is the average response time of the REST APIs of Twitter that returns the information of a user given his/her ID. For more information see https://dev.twitter.com/rest/reference/get/users/lookup.

[2] http://www.w3.org/TR/sparql11-federated-query/.

To answer this research question, first, *Filter Update Policy* is proposed for refreshing the local replica. Then, the extension of ACQUA policies is introduced by combining them with the proposed policy, named *ACQUA.F*. The experimental evaluations demonstrate their efficiency comparing their performance against those of the ACQUA policies.

The remainder of the chapter is organized as follows. Section 4.2 formalizes the problem. Section 4.3 introduces the proposed solution for refreshing the replica of distributed data and discusses *ACQUA.F* policies in detail. Section 4.4 provides experimental evaluations for investigating the hypotheses, and, finally, Sect. 4.5 concludes.

4.2 Problem Statement

In order to investigate the research question, this study considers continuous RSP-QL queries over a data stream S and a distributed data B under the following two assumptions: (*i*) There is a 1:1 join relationship between the data items in the data stream and those in the distributed data; and (*ii*) The distributed data is evolving and it slowly changes between two subsequent evaluations. In particular, the study considers the class of queries with the algebraic expression SE of the following form:

$$(WINDOW \; u_S \; P_S) \; JOIN \; ((SERVICE \; u_B \; P_B) \; FILTER \; F), \qquad (4.1)$$

where: (*i*) P_S, and P_B are basic graph patterns, (*ii*) u_S, and u_B identify the window on the RDF stream and the remote SPARQL endpoint, and (*iii*) F is either ($?x < \mathcal{FT}$) or ($?x > \mathcal{FT}$), where $?x$ is a variable in P_B and $\mathcal{FT}$ is the *Filtering Threshold*.

Let Ω^W be the set of solution mappings returned from the WINDOW clause, Ω^S be the one returned from the SERVICE clause and Ω^R be the one stored in the replica. Applying a Filtering Threshold $\mathcal{FT}$ to a variable $?x$ that appears in Ω^S means that for each mapping $\mu^S \in \Omega^S$ it checks $\mu^S(?x) > \mathcal{FT}$, and discards the mappings which do not satisfy the condition.

4.3 Proposed Solution

This section introduces the proposed solution. Section 4.3.1 discusses the proposed *Filter Update Policy* as a ranker for the maintenance process of the local replica $\mathcal{R}$. Section 4.3.2 shows how the proposed approach can improve the ACQUA policies by integrating them with *Filter Update Policy*.

Algorithm 4.1: The pseudo-code of the Filter Update Policy

---

1 **foreach** $\mu^{\mathcal{R}} \in \mathcal{C}$ **do**
2 $\quad \mid \quad FD(\mu^{\mathcal{R}}) = |\mu^{\mathcal{R}}(?x) - \mathcal{FT}|$;
3 **end**
4 order $\mathcal{C}$ w.r.t. the value of $FD(\mu^{\mathcal{R}})$;
5 $\mathcal{E}$ = first γ mappings of $\mathcal{F}$;
6 **foreach** $\mu^{\mathcal{R}} \in \mathcal{E}$ **do**
7 $\quad \mid \quad \mu^S = \text{ServiceOp.next(JoinVars}(\mu^{\mathcal{R}}))$;
8 $\quad \mid \quad$ replace $\mu^{\mathcal{R}}$ with μ^S in $\mathcal{R}$;
9 **end**

4.3.1 Filter Update Policy

This section introduces **Filter Update Policy** for refreshing the local replica $\mathcal{R}$ of the distributed data. As already stated in Sect. 4.2, a class of continuous SPARQL queries is considered that join the streaming data with distributed data and the SERVICE clause contains a FILTER condition.

The result of the SERVICE clause is stored in the replica $\mathcal{R}$. The maintenance process introduced in Chap. 3 consists of the following components: the proposer, the ranker, and the maintainer. This solution exploits **WSJ** algorithm from ACQUA as a proposer, i.e., it selects the set $\mathcal{C}$ of candidate mappings for the maintenance from the current window. The **Filter Update Policy** computes the elected set $\mathcal{E} \subseteq \mathcal{C}$ of mappings to be refreshed as a ranker and, finally, the maintainer refreshes the mappings in set $\mathcal{E}$.

For each mapping in the replica defined as $\mu^{\mathcal{R}}$, **Filter Update Policy** (*i*) computes how close is the value associated to the variable $?x$ in the mapping $\mu^{\mathcal{R}}$ to the Filtering Threshold $\mathcal{FT}$, and (*ii*) selects the top γ ones for refreshing the replica (where γ is the refresh budget). In order to compute the distance between the value of $?x$ in mapping $\mu^{\mathcal{R}}$ and Filtering Threshold $\mathcal{FT}$, this study defines the *Filtering Distance* $\mathcal{FD}$ of mapping $\mu^{\mathcal{R}}$:

$$FD(\mu^{\mathcal{R}}) = |\mu^{\mathcal{R}}(?x) - \mathcal{FT}| \tag{4.2}$$

If the value associated to $?x$ smoothly changes over time,[3] then, intuitively, the smaller the Filtering Distance of a mapping in the last evaluation, the higher is the probability to cross the Filtering Threshold $\mathcal{FT}$ in the current evaluation and, thus, to affect the query evaluation. For instance in Listing 4.1, for each user the Filtering Distance between the number of followers and the Filtering Threshold $\mathcal{FT} = 100,000$ is computed. Users, whose numbers of followers were closer to 100,000 in the last evaluation, are more likely to affect the current query evaluation.

---

[3]The wording *smoothly changes over time* means if $?x = 98$ in the previous evaluation and $?x = 101$ in the current evaluation, in next evaluation it is more likely that $?x = 99$ than jumping to $?x = 1000$.

Algorithm 4.2: The pseudo-code of integrating Filter Update Policy with ACQUA's ones

---

1 **foreach** $\mu^{\mathcal{R}} \in \mathcal{C}$ **do**
2     $FD(\mu^{\mathcal{R}}) = |\mu^{\mathcal{R}}(?x) - \mathcal{FT}|$
3   **if** $FD(\mu^{\mathcal{R}}) < \mathcal{FDT}$ **then**
4 | add $\mu^{\mathcal{R}}$ to $\mathcal{F}$;
5   **end**
6 **end**
7 UP ($\mathcal{F}$, γ, update policy);

Algorithm 4.1 shows the pseudo-code of the **Filter Update Policy**. For each mapping in the candidate mapping set $\mathcal{C}$, it computes the Filtering Distance as the absolute difference of the value $?x$ of mapping $\mu^{\mathcal{R}}$ and the Filtering Threshold $\mathcal{FT}$ in the query (Lines 1–3). Then, it orders the set $\mathcal{C}$ based on the absolute differences (Line 4). The set of elected mapping $\mathcal{E}$ is created by getting the top γ ones from the ordered set of $\mathcal{F}$ (Line 5). Finally, the local replica $\mathcal{R}$ is maintained by invoking the SERVICE operator and querying the remote SPARQL endpoint to get fresh mappings and replace them in $\mathcal{R}$ (Lines 6–9).

4.3.2 ACQUA.F Policies

It is worth to note that **Filter Update Policy** can be combined with the policies proposed in ACQUA. The intuition is simple, it is useless to refresh data items that are not likely to pass the filter condition; it is better to focus on a *band* around the condition.

In this new approach, first, the proposer, using **WSJ** algorithm, generates the candidate set $\mathcal{C}$, then the **Filter Update Policy** determines the data items that fall in the band, and, then, applies one of the ACQUA policies select among those data items a set of mappings $\mathcal{E} \subset \mathcal{C}$ to be refreshed in replica $\mathcal{R}$ as a ranker. Finally, the maintainer updates the replica. The proposed policies are collectively named **ACQUA.F**.

Algorithm 4.2 shows the pseudo-code that integrates the **Filter Update Policy** with ACQUA ones. It is worth to note that this algorithm requires to get the value of *band* as a parameter, namely *Filtering Distance Threshold* $\mathcal{FDT}$. In this chapter, there is an assumption that $\mathcal{FDT}$ is easy to be determined and the proposed policies are not sensitive to the predefined value of $\mathcal{FDT}$. In the next chapter, this assumption will be relaxed. For each mapping in the candidate mapping set $\mathcal{C}$, it computes the Filtering Distance (Lines 1–2). If the difference is smaller than Filtering Distance Threshold $\mathcal{FDT}$, it adds the mapping to the set $\mathcal{F}$ (Lines 3–5). Given the set $\mathcal{F}$, the refresh budget γ, and the update policy name (WSJ-RND, WSJ-LRU, and WSJ-WBM), the function UP considers the set $\mathcal{F}$ as the candidate set and applies the named policy on it (Line 7). The three adapted policies respectively are named RND.F, LRU.F, and WBM.F.

4.4 Experiments

This section reports the result of experiments that evaluate the proposed policies. Section 4.4.1 introduces hypotheses. Section 4.4.2 introduces the experimental setting that is used to check the validity of the hypotheses. Section 4.4.3 discusses the experiments related to the first hypothesis and shows the related result. Finally, Sect. 4.4.4 shows the results related to the second hypothesis.

4.4.1 Hypotheses

To answer the research question presented in Sect. 4.1, the following two hypotheses are formulated:

Hp.2.1 The replica can be maintained fresher than when using ACQUA policies by first refreshing the mappings $\mu^{\mathcal{R}} \in \Omega^{\mathcal{R}}$ for which $\mu^{\mathcal{R}}(?x)$ is closer to the Filtering Threshold.

Hp.2.2 The replica can be maintained fresher than when using ACQUA policies by first selecting the mappings as in Hypothesis Hp.2.1 and, then, applying the ACQUA policies.

4.4.2 Experimental Setting

As an experimental environment, an Intel i7 @ 1.7 GHz with 4 GB memory and an hdd disk are used. The operating system is Mac OS X Lion 10.9.5 and Java 1.7.0.67 is installed on the machine. The experiments are done by extending the experimental setting of ACQUA (see also Sect. 3.4.1).

The experimental datasets are composed of streaming and distributed data. The streaming data is collected from 400 verified users of Twitter for three hours of tweets using the streaming API of Twitter. The distributed data is collected invoking the Twitter API, which returns the number of followers per user, every minute during the three hours of recording the streaming data. As a result, for each user, the distributed data contain a time-series that records the number of followers.

In order to control the selectivity of the filtering condition, a transformation of the distributed data is designed that randomly selected a specified percentage of the users (i.e., 10, 20, 25, 30, 40 and 50%) and, for each user, translates the time-series, which captures the evolution overtime of the number of followers, to be sure that it crosses the Filtering Threshold[4] at least once during the experiment. In particular, for each user, first, the minimum and maximum number of followers are found; then, the difference of the minimum number of followers and Filtering Threshold is defined as *MaxDifference*. The difference of the maximum number of followers and Filtering Threshold also is defined as *MinDifference*. Finally, a number between *MinDifference* and *MaxDifference* is generated randomly and added to each value of the time-series of the number of followers of the selected user.

It is worth to note that this translation does not alter the nature of the evolution over time of the number of followers, it only moves the entire time-series so that it crosses the Filtering Threshold at least once during the experiment. If the original time-series of the number of followers is almost flat (i.e., it slightly moves up and down around a median) or it is fast growing/shrinking; then the translated time-series will have the same trend. The only effect of the translation is to control the selectivity of the filter operator. So, the experiments are limited to six values of the selectivity computed as (100-percentage) (90, 80, 75, 70, 60, and 50%).

In order to reduce the risk to introduce a bias in performing the translation, the procedure is repeated 10 times for each selectivity listed above, generating 10 different datasets for each selectivity. Each group of 10 datasets using specific selectivity is named *test case* and is labeled with DSx%; for example, DS10% test case identifies the 10 datasets in which the number of followers of 10% of the users crosses the Filtering Threshold at least once during the experiment.

The query presented in Sect. 4.1 is used for experiment and ran for 140 iterations of the query evaluation. In order to investigate the hypotheses, an Oracle is organized that, at each iteration i, certainly provides corrects answers $Ans(Oracle_i)$ and its answers can be compared with the possibly erroneous ones of the query $Ans(Q_i)$. Given that the answer to the query in Listing 4.1 is a set of users' IDs, Jaccard distance is used to measure diversity of the set generated by the query and the one generated by the Oracle, and cumulative Jaccard distance at the kth iteration $d_J^C(k)$ is introduced as:

$$d_J^C(k) = \sum_{i=1}^{k} d_J(Ans(Q_i), Ans(Oracle_i)) \tag{4.3}$$

where $d_J(Ans(Q_i), Ans(Oracle_i))$ is the Jaccard distance of iteration i.

4.4.3 Experiment 1—Sensitivity for Filter Update Policy

This experiment investigates the first hypothesis (Hp.2.1). In order to verify the hypothesis, the proposed policy is compared with ACQUA ones. The worst maintenance policy is WST which does not refresh the replica $\mathcal{R}$ during the evaluation and,

---

[4]The value of the Filtering Threshold is chosen to guarantee that no one of the original time-series crosses it.

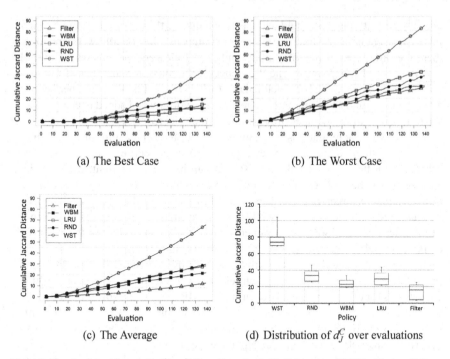

(a) The Best Case

(b) The Worst Case

(c) The Average

(d) Distribution of d_J^C over evaluations

Fig. 4.1 Result of experiment that investigates Hypothesis Hp.2.1 testing the Filter Update Policy and the state-of-the-art policies proposed in ACQUA over DS75% test case

thus, is an upper bound for the error. WSJ is used as a proposer for all maintenance policies. As described in Chap. 3, WSJ selects the mappings from the ones currently involved in the evaluation and creates the candidate set $\mathcal{C}$. For ranker WSJ-RND, WSJ-LRU, and WSJ-WBM policies are used, which are introduced in Chap. 3. WSJ-RND update policy randomly selects the mappings while WSJ-LRU chooses the least recently refreshed mappings. WSJ-WBM identifies the possibly stale mappings and chooses them for maintenance.

It is important to note that there are two points of view to show the results of the experiments. The first point of view takes a time-series perspective and compare various policies through the time for every evaluation. The second point of view focuses on the distribution of the cumulative Jaccard distance at the end of the experiment, and uses a box-plot [6] to highlight the median and the four quartiles of the accuracy obtained from the experiments.

In this experiment, the refresh budget γ is equal to 3, and the DS75% test case (where the numbers of followers of 25% of the users cross the Filtering Threshold) is selected. The experiment runs 140 iterations of query evaluation over each of the 10 different distributed datasets. Figure 4.1 shows the result of the experiment. Figures 4.1a and b respectively show the best and the worst runs using the first viewpoints. Figure 4.1c presents the average of the results obtained with the 10 datasets.

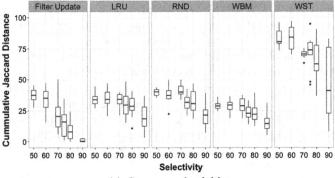

(a) Compare selectivities

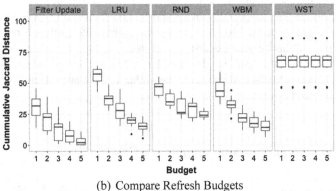

(b) Compare Refresh Budgets

Fig. 4.2 Result of experiment that investigates how the result presented in Fig. 4.1 changes using different selectivities and refresh budgets

As the result shows, the Filter Update Policy is the best one in all cases. The WSJ-WBM is better than the WSJ-RND and the WSJ-LRU on average and in the worst case, but the WSJ-LRU is better than WSJ-WBM in the best case. As expected, the WST policy is always the worst one.

Figure 4.1d which uses the second viewpoint, shows the distribution of cumulative Jaccard distance at the 140th iteration obtained with the DS75% test case. As the result shows, the Filter Update Policy outperforms other policies in 50% of the cases. Comparing the WSJ-WBM policy with WSJ-RND and WSJ-LRU policies, WSJ-WBM performs better than WSJ-RND in 50% of the cases. The WSJ-LRU policy also performs better than WSJ-RND in average. As expected, the WST policy has always the highest cumulative Jaccard distance.

To check the sensitivity to the filter selectivity, in the evaluated case, to the percentage of users whose number of followers is crossing the filtering threshold, the experiment is repeated with different datasets in which the selectivity is changed. Keeping the refresh budget γ equal to 3, this study runs different experiments with the test cases DS90%, DS80%, DS70%, DS60%, DS50%. As for the DS75% test case, this study generates 10 datasets for each value of selectivity and runs the experiment on them. For each dataset and each policy the median, the first quartile, and the third quartile of cumulative Jaccard distance at the 140th iteration over 10 datasets are computed. Figure 4.2a shows the obtained results. The Filter Update Policy has better performance than the other ones for DS90%, DS80%, DS75% and DS70% test cases. Intuitively, in those datasets, there exist fewer users whose number of followers crosses the Filtering Threshold, so the correct user for updating is selected with higher probability. The result also shows that the behavior of WSJ-WBM policy is stable over different selectivities and performs better than Filter Update Policy over datasets DS60% and DS50% test cases.

In order to check the sensitivity to the refresh budget, the experiment is repeated with different refresh budgets (γ equals to 1, 2, 3, 4, and 5) over DS75% test case. Figure 4.2b shows the median, the first quartile, and the third quartile of cumulative Jaccard distance at the 140th iteration over test cases for different policies and budgets. The cumulative Jaccard distance in WST does not change for different budgets, but for all other policies the cumulative Jaccard distance decreases when the refresh budget increases; this means that higher refresh budgets always leads to fresher replica and fewer errors.

4.4.4 Experiment 2—Sensitivity for ACQUA.F Policies

This experiment investigates the second hypothesis (Hp.2.2). Using DS25% test case, the performances of RND.F, LRU.F, and WBM.F, which respectively combine the Filter Update Policy with WSJ-RND, WSJ-LRU, and WSJ-WBM, compared with state of the art ones. Assuming that determining the band around the Filter Threshold a priori is simple, the Filtering Distance Threshold $\mathcal{FDT}$ parameter is set to 1,000 (1% of the maximum number of followers).[5] As explained in Sect. 4.3, those new policies, first, create the candidate set $\mathcal{C}$, then they reduce the candidate set by omitting the users that have Distance Threshold greater than $\mathcal{FDT}$ and, finally, they apply the rest of the ACQUA policy to the candidate set which selects the mappings for refreshing in the replica $\mathcal{R}$.

Figure 4.3 shows the result of the experiment. In Fig. 4.3a, the chart shows the cumulative Jaccard distance across the 140 iterations in the best run. In this case, the Filter Update Policy performs better in most of the iterations. Figure 4.3b shows the worst case, where the LRU.F policy is the best one in all the iterations. Figure 4.3c shows the average performance of the policies. The LRU.F policy is the best one also in this case. Figure 4.3d shows the distribution of the cumulative Jaccard distance over DS75% test case. The LRU.F policy performs better than RND.F and WBM.F in most of the cases. The WBM.F policy performs better than RND.F policy in most of the cases.

---

[5]Later, in the next chapter, it is found that determining the band a priori is not straightforward.

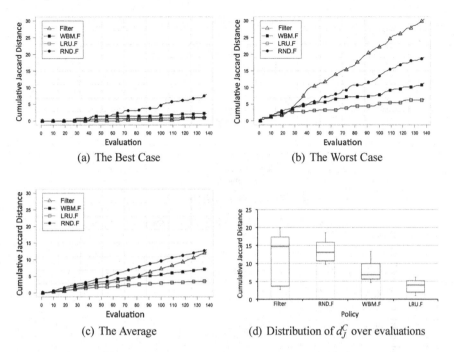

Fig. 4.3 Result of experiment that compares Filter Update Policy with ACQUA.F policies to investigate Hp.2.2

To check the sensitivity to the filter selectivity, the experiment is repeated with different datasets in which this selectivity is changed. The experiments run over the DS90%, DS80%, DS75%, DS70%, DS60%, and DS50% test cases, while keeping the refresh budget γ equal to 3. For each test case and each policy, the median, the first quartile, and the third quartile of cumulative Jaccard distance at the last iteration over the 10 datasets are computed. Figure 4.4a shows the obtained results. The LRU.F policy always has better performance than the other ones. The behavior of LRU.F, and WBM.F policies are stable over different selectivities. The Filter Update Policy has better performance than WBM.F for DS90%, DS80%, DS75% and DS70% test cases. In those datasets, there exist fewer users whose number of followers crosses the Filtering Threshold, and with a higher probability the correct user for updating is selected.

The experiment is repeated with different refresh budgets to check the sensitivity of the result to the refresh budget for ACQUA.F policies (γ equals to 1, 2, 3, 4, and 5). The experiment run over the DS75% test case. Figure 4.4b shows the median, the first quartile, and the third quartile of cumulative Jaccard distance at the last iterations over 10 datasets for different policies and budgets. The cumulative Jaccard distance in WST does not change for different budgets, but for all other policies when the refresh budget increases, the cumulative Jaccard distance decreases which means that higher refresh budgets always leads to a fresher replica and fewer errors.

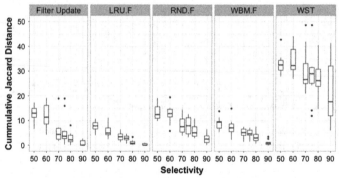

(a) Compare selectivities

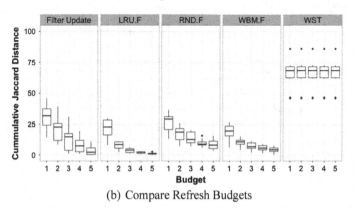

(b) Compare Refresh Budgets

Fig. 4.4 Result of experiment that investigates how the result presented in Fig. 4.3 changes using different selectivities and refresh budgets

4.5 Outlook

This chapter studied the problem of the continuous evaluation of a class of queries that joins streaming and distributed data and includes a FILTER condition in the SERVICE clause. Reactiveness is the most important performance indicator for evaluating queries in RSP engines. When the data is distributed over the Web and slowly evolves over time (i.e., it is quasi-static), correct answers may not be reactive, because the time to access the distributed data may exceed the time between two consecutive evaluations.

To address this problem, ACQUA (presented in Chap. 3) proposes to use (*i*) a replica to store the distributed data at query registration time, (*ii*) a maintenance policy to keep the data in the replica fresh and (*iii*) a refresh budget to limit the number of accesses to the distributed data. In this way, accurate answers can be provided while meeting operational deadlines.

Table 4.1 Summary of the verification of the hypotheses w.r.t. Filter Update Policy, LRU.F, WBM.F, and RND.F

	Measuring	Varying	Filter Update Policy	LRU.F	WBM.F	RND.F
Hp.2.1	accuracy	selectivity	>60%			
Hp.2.1	accuracy	budget	✓			
Hp.2.2	accuracy	selectivity		✓		
Hp.2.2	accuracy	budget		>2	=1	

This chapter extends ACQUA, and in particular, investigates queries where (*i*) the algebraic expression is a FILTER of a JOIN of a WINDOW and a SERVICE and (*ii*) the filter condition selects mappings from the SERVICE clause checking if the values of a variable are larger (or smaller) than a Filtering Threshold.

To study this class of queries, two hypotheses are formulated that capture the same intuition: the closer was the value to the Filtering Threshold in the last evaluation, the more probable is that it will cross the Filtering Threshold in the current evaluation, and, thus, it is a mapping to refresh. In Hypothesis Hp.2.1, this intuition is tested by defining the new Filtering Update Policy, whereas, in Hypothesis Hp.2.2, the intuition is tested together with the ACQUA policies defining RND.F, LRU.F, and WBM.F respectively extending WSJ-RND, WSJ-LRU and, WSJ-WBM. The results are reported in Table 4.1.

The result of experiments about Hp.2.1 shows that Filter Update Policy keeps the replica fresher than ACQUA policies when the selectivity of filtering condition is above 60% of the total. Below this selectivity ACQUA results are confirmed and WSJ-WBM remains the best choice.

The results of the experiments about Hp.2.2 shows that the Filter Update Policy can be combined with ACQUA policies in order to keep the replica even fresher than with the Filter Update Policy.

References

1. Barbieri DF, Braga D, Ceri S, Valle ED, Grossniklaus M (2010) Querying rdf streams with c-sparql. ACM SIGMOD Record 39(1):20–26
2. Calbimonte JP, Jeung H, Corcho O, Aberer K (2012) Enabling query technologies for the semantic sensor web. Int J Semant Web Inf Syst 8:43–63
3. Dell'Aglio D, Calbimonte JP, Della Valle E, Corcho O (2015) Towards a unified language for rdf stream query processing. In: International semantic web conference. Springer, Cham, pp 353–363
4. Dell'Aglio D, Della Valle E, Calbimonte JP, Corcho O (2014) RSP-QL semantics: a unifying query model to explain heterogeneity of RDF stream processing systems. Int J Semant Web Inf Syst 10(4):17–44

5. Le-Phuoc D, Dao-Tran M, Parreira JX, Hauswirth M (2011) A native and adaptive approach for unified processing of linked streams and linked data. In: The Semantic Web–ISWC 2011. Springer, Berlin, pp 370–388
6. McGill R et al (1978) Variations of box plots. Am Statist 32(1):12–16

Chapter 5
Rank Aggregation in Queries with a FILTER Clause

Abstract This chapter focuses on using rank aggregation in continuous query answering over streaming and distributed data in the RSP context. When data is distributed over the Web, the time to access distributed data can put the application at risk of losing reactiveness. Keeping a local replica of the distributed data lets the RSP engines remain reactive. In previous chapters, different maintenance policies are investigated which guarantee reactiveness while maximizing the freshness of the replica. Those policies simplify the problem with several assumptions. This chapter focuses on removing some of those simplification assumptions. In particular, the proposed solution shows that rank aggregation can be effectively used to fairly consider the opinions of different policies. The experimental evaluation illustrates that rank aggregation is key to move a step forward to the practical solution of this problem.

Keywords Continuous query answering · Rank aggregation · Streaming data · Evolving distributed data · RDF data · RSP engine

5.1 Introduction

Being reactive[1] is the most important requirement in application domains that combine data stream with distributed data. Consider Example 3 in Chap. 1: an advertising agency may want to propose viral contents to influencers (e.g., users with more than 100,000 followers) when they are mentioned in micro-posts across Social Networks. In this case, query respond is needed in a timely fashion, because (*i*) followers have few minutes of the attention span, and (*ii*) competitors may try to reach those influencers before us.

However, the time to access and fetch the distributed data can be so high to put the application at risk of losing reactiveness. This is especially true when the distributed data evolves; e.g., in Example 1 in Chap. 1, the number of followers of the influencers,

---

[1] A program is reactive if it maintains a continuous interaction with its environment, but at a speed which is determined by the environment, not by the program itself [2]. Real-time programs are reactive, but reactive programs can be non real-time as far as they provide result in time to successfully interact with the environment.

© The Author(s), under exclusive license to Springer Nature Switzerland AG 2020   55
S. Zahmatkesh and E. Della Valle, *Relevant Query Answering over Streaming and Distributed Data*, PoliMI SpringerBriefs,
https://doi.org/10.1007/978-3-030-38339-8_5

which are in the distributed data, is more likely to change when they are mentioned in the stream.

Although RDF Stream Processing (RSP) engines [5] are an adequate framework to develop this type of applications, they may lose reactiveness when accessing data distributed over the Web.

As stated in Chap. 4, ACQUA policies cannot fully optimize queries like the one in Listing 4.1, because they do not consider the FILTER clause. The previous chapter introduces the extension of ACQUA to optimize the class of queries that include the filtering of the intermediate results obtained from the federated SPARQL endpoint. The intuition is simple, it is useless to refresh data items that are not likely to pass the filter condition; it is better to focus on a *band* around the condition. For instance, for the query in Listing 4.1, the engine focuses on the band $?fCount \in [90000, 110000]$. This new approach first determines the data items that fall in the band (namely, the Filter Update Policy) and, then, applies one of the ACQUA policies on those data items. The proposed policies are collectively named ACQUA.F. The result of the evaluation shows that ACQUA.F policies outperform ACQUA ones, when the selectivity of the FILTER clause is high.

ACQUA.F approach assumes that determining a priori the band to focus on is simple. This chapter further investigates new approaches by removing this assumption, because experimental evidence shows that it is difficult to determine such a band. Instead of applying in a pipe the Filter Update Policy and one of the ACQUA policies, each policy can express its *opinion* by ranking data items according to its criterion and, then, rank aggregation [8] takes all opinions fairly into account. For this reason the research question is:

RQ.3   *Can rank aggregation combine the ACQUA policies with* Filter Update Policy, *so to continuously answer queries (such as the one in Listing 4.1) and to guarantee reactiveness while keeping the replica fresh (i.e., giving results with high accuracy)?*

In particular, the contributions of this chapter are the following:

- Providing empirical evidence that relaxing the ACQUA.F assumption is hard, i.e., it is hard to determine a priori the band to focus on.
- Defining three new policies (collectively named, ACQUA.F$^+$) that use rank aggregation to combine the Filter Update Policy and the ACQUA policies.
- Demonstrating empirically on synthetic and real datasets that one of the new ACQUA.F$^+$ policies keep the replica as fresh as the corresponding ACQUA.F one but without requiring to determine a priori the band to focus on.
- Demonstrating empirically that such a policy uses the same budget as the corresponding ACQUA.F policy.

The remainder of the chapter is organized as follows. Section 5.2 introduces the rank aggregation method. Section 5.3 introduces the idea of using rank aggregation to combine the ACQUA policies with Filter Update Policy in order to reactively answer continuous queries while keeping the replica fresh (i.e., giving results with high accuracy). The proposed rank aggregation solution is introduced in Sect. 5.4. Section 5.5 details the research hypotheses, introduces the experimental settings, reports on the evaluation of the proposed methods and discusses the gathered practical insight. Section 5.6 concludes wrapping up the achievements.

5.2 Rank Aggregation

In many circumstances, there is the need to rank a list of alternative options (namely, candidates) according to multiple criteria. For instance, in many sports, the ranking of the athletes is based on the individual scores given by several judges. The problem of computing a single rank, which fairly reflects the opinion of many judges, is called *rank aggregation* [8]. Rank aggregation algorithms get m ranked lists and combine them to produce a single ranking which describes the preferences in the given m lists in the best way. This section introduces different rank aggregation metrics include consensus-based ranking (Borda count [3]), and pairwise disagreement based ranking (Kemeny optimal aggregation [8]).

Borda's method [3] is an election method in which voters rank candidates in order of preference. In this method, each candidate gets a score based on its position in the list. In Borda's method, assuming that there are V voters, each identifies a set C of candidates with a preference order. First, each of the candidates in the position n, is given $|C| - n$ as score, so, for each candidate c and each voter v, the score computed as $score(c, v) = |C| - n$, where the position of candidate c in the vote of voter v is equal to n. Then, the candidates are ranked by their total score, e.g., by the weighted sum of the scores given by the individual voters. This method was aimed for use in elections with a single winner, but it is also possible to use it for more than one winner, by defining the top k candidate with the highest scores as the winners.

In pairwise disagreement based algorithms, the metric which measures the distance between two ordered lists should be optimized.

Definition 5.1 *Ordered List.* Given a universe U, an ordered list with respect to U is an ordering of a subset $S \in U$, i.e., $\tau = [x_1 \geq x_2 \geq \ldots \geq x_d]$, where $x_i \in S$, and $\geq$ is an ordering relation on S. $\tau(i)$ denotes the position or rank of i.

The list τ is a *full list* if it contains all the elements in the universe U, otherwise it is named *partial list*. There is also a special case named *top k list*, where only ranks a subset of S, and all the ranked elements are above the unranked ones. The size of the ranked list is equal to k. There exist two popular metrics to measure the distance between two full lists [6]:

- The *Spearman footrule distance* is the sum of the absolute difference between the rank of element i according to the two lists, for all $i \in S$, i.e., $F(\tau, \sigma) = \sum_1^{|S|} |\tau(i) - \sigma(i)|$.

- The *Kendall tau distance* counts the number of pairwise disagreements between two lists. The distance between two full lists τ and σ is equal to $K(\tau, \sigma) = |(i, j)| \, i < j$, and $\tau(i) < \tau(j)$; but $\sigma(i) > \sigma(j)|$.

The aggregation obtained by optimizing Kendall distance is called *Kemeny optimal aggregation*, while the one optimizing the Spearman footrule distance is called *footrule optimal aggregation*. Dwork et al. [8] show that *Kemeny optimal aggregation* is NP-Hard, but it is possible to approximate the Kendall distance via Spearman footrule distance, and footrule optimal aggregation has polynomial complexity.

In addition to the mathematical perspective in rank aggregation studies [3, 4, 12, 13], there exist various researches in different fields such as database community [1, 9–11].

As stated in Sect. 2.4, Fagin et al. [11] introduce "Threshold Algorithm" or TA, for aggregating different ranked lists of objects based on different criteria (i.e. objects' attributes) to determine the top k objects.

Bansal et al. [1] address the problem of object clustering. Objects are represented as a vertex of a graph with edge labeled $(+)$ or $(-)$ for each pair of objects, indicating that two objects should be in the same or different clusters, respectively. The goal is to cluster the objects in a way that minimizes the edges with $(-)$ label inside clusters and edges with $(+)$ label between clusters, which is known as correlation-clustering.

In addition to the metrics introduced in [8] for aggregation of fully ranked list, Fagin et al. [10] introduce various metrics for top-k list, based on various motivating criteria. Getting the idea from [7], they also propose the notion of an equivalence class of distance measures as follows: Two distance measures d and d' are equivalent if there are positive constants c_1 and c_2 such that $c_1 d'(\tau_1, \tau_2) \leq d(\tau_1, \tau_2) \leq c_2 d'(\tau_1, \tau_2)$ for every pair τ_1, τ_2 of top k lists. They show that many of the proposed distance measures can fit into one large equivalence class.

Later, Fagin et al. [9] introduce four different metrics for partially ranked lists. They extend the Kendall tau distance and the Spearman footrule distance using different approaches, and prove that their metrics are equivalent.

In this chapter rank aggregation is selected because the problem addressed in this study requires to minimize the time spent in any computation and Borda's method is computationally easy. A naïve algorithm can solve the rank aggregation problem using Borda's method in linear time and algorithms exist that can solve it in sub-linear time, e.g. the Threshold Algorithm (Sect. 2.4). Other methods exist to handle cases where not all the voters can give a score to all the candidates or the case where some voters are biased or even malicious. However, those methods are computationally more expensive and handle problems that do not appear in our rank aggregation scenario.

5.3 Problem Statement

ACQUA.F applies the Filter Update Policy and the ACQUA policies in a pipe. In this way, the *opinion* of the Filter Update Policy is more relevant than the one of ACQUA policies. This gives good result when focusing on a band around the $\mathcal{FT}$ minimizes the number of stale data. So, ACQUA.F assumes that it is possible to determine a priori the band to focus on, i.e., the optimal value of the Filtering Distance Threshold. However, when the selectivity of the filter condition is low, focusing on such a band is inconvenient. Later this chapter shows that relaxing this assumption is hard.

Rank aggregation [8] was shown to be an adequate solution in similar settings where there was the need to take fairly into account the opinions of different algorithms. Therefore, it is considered as an alternative solution for combining the maintenance policies.

As in the previous chapter, the class of queries where the algebraic expression SE is in the following form is considered:

$$(WINDOW \; u_{\mathcal{S}} \; P_{\mathcal{S}}) \; JOIN \; ((SERVICE \; u_{\mathcal{B}} \; P_{\mathcal{B}}) \; FILTER \; F), \qquad (5.1)$$

where $P_{\mathcal{S}}$, and $P_{\mathcal{B}}$ are graph patterns, $u_{\mathcal{S}}$, and $u_{\mathcal{B}}$ identify the window on the RDF stream and the remote SPARQL endpoint, and F is either ($?x < \mathcal{FT}$) or ($?x > \mathcal{FT}$), where $?x$ is a variable in $P_{\mathcal{B}}$ and $\mathcal{FT}$ is the *Filtering Threshold*.

In the proposed solution, WSJ is used as proposer to select the candidate set $\mathcal{C}$ of mappings for the maintenance. As a ranker, rank aggregation is used to combine the ranking obtained by ordering the mappings in $\mathcal{C}$ according to the scores computed by each policy. Specifically, a weight (denoted with α) allows computing an aggregated score as follows:

$$score_{Agg} = \alpha * score_{Filter} + (1 - \alpha) * score_{ACQUA} \qquad (5.2)$$

The aggregated list is ordered by the score $score_{Agg}$. In the next step, the maintenance process selects the top γ ones from the ordered list to create the elected set $\mathcal{E} \subseteq \mathcal{C}$ of mappings to be refreshed. Finally, the maintainer refreshes the mappings in set $\mathcal{E}$.

5.4 Rank Aggregation Solution

This section introduces the proposed solution to the problem of combining in a timely fashion data stream with distributed data in the context of RSP. Section 5.4.1 shows how to apply the idea of rank aggregation for combining the WSJ-LRU and WSJ-WBM policies with the Filter Update Policy to obtain LRU.F$^+$, and WBM.F$^+$, respectively. Section 5.4.2 introduces a different method to combine WSJ-WBM and the Filter Update Policy to obtain WBM.F*.

5.4.1 ACQUA.F⁺ Policy

This section presents an algorithm to combine Filter Update policy with ACQUA policies, respectively, named LRU.F⁺, and WBM.F⁺.

In this new combined approach, the proposer selects a set $\mathcal{C}$ of candidate mappings for the maintenance, then the proposed policy receives as input the parameter α and the two ranked lists of mappings $\mathcal{CL}$ generated by ACQUA policy and Filter Update Policy, and it generates a single ranked list of mappings.

Algorithm 5.1 shows the pseudo-code of the proposed policy. For each mapping in the candidate set $\mathcal{C}$ it computes the Filtering Distance as the absolute difference of the value $?x$ of mapping $\mu^{\mathcal{R}}$ and the Filtering Threshold $\mathcal{FT}$ in the query (Lines 1–3). Then, it orders the set $\mathcal{C}$ based on the Filtering Distance and generates the ranked list $\mathcal{CL}_f$ (Line 4). In the next step, based on the selected policy from ACQUA, the algorithm computes the score for each mapping in the candidate set $\mathcal{C}$ (Lines 5–7), and orders the candidate set based on the scores to generate the ranked list $\mathcal{CL}_{acqua}$ (Line 8).

For LRU.F⁺ policy, the algorithm computes the refresh time for each mapping in the candidate set $\mathcal{C}$, and generates scores based on the least recently refreshed mappings. For WBM.F⁺ policy, for each mapping in the candidate set $\mathcal{C}$, the remaining life time, the renewed best before time, and the final score are computed to order the candidate set.

Given the parameter α and the two ranked lists $\mathcal{CL}_f$ and $\mathcal{CL}_{acqua}$, the Function AggregateRanks generates a single ranked list $\mathcal{CL}_{agg}$ aggregating the scores of two lists (Line 9). The set of elected mappings $\mathcal{E}$ is created by getting the top γ ones from $\mathcal{CL}_{agg}$ (Line 10). Finally, the local replica $\mathcal{R}$ is maintained by invoking the SERVICE operator and querying the remote SPARQL endpoint to get fresh mappings and replace them in the replica $\mathcal{R}$ (Lines 11–14).

5.4.2 WBM.F* Policy

This section introduces WBM.F*, an improved version of WBM.F⁺. It considers that the candidate set $\mathcal{C}$ in WSJ-WBM algorithm has two subsets that include the "Expired" and the "Not Expired" mappings, respectively. WSJ-WBM policy uses the refresh budget only to update the mappings from the "Expired" set.

The proposed WBM.F* algorithm computes the "Expired" and "Not Expired" lists of WSJ-WBM policy, and accordingly, generates two ranked lists ordering them based on Filter Update Policy. Finally, using rank aggregation, WBM.F* generates two ranked lists, "Expired.agg" and "Not Expired.agg". WBM.F* policy first selects

---

Algorithm 5.1: The pseudo-code of the **ACQUA.F⁺**

---

1 **foreach** $\mu^{\mathcal{R}} \in \mathcal{C}$ **do**
2     $FD(\mu^{\mathcal{R}}) = |\mu^{\mathcal{R}}(?x) - \mathcal{FT}|$;
3 **end**
4 $\mathcal{CL}_f$ = order $\mathcal{C}$ w.r.t. the value of $FD(\mu^{\mathcal{R}})$;
5 **foreach** $\mu^{\mathcal{R}} \in \mathcal{C}$ **do**
6   compute the score of $\mu^{\mathcal{R}}$ based on the selected policy from ACQUA ;
7 **end**
8 $\mathcal{CL}_{acqua}$ = order $\mathcal{C}$ w.r.t. the generated scores;
9 $\mathcal{CL}_{agg} = AggregateRanks(\alpha, \mathcal{CL}_f, \mathcal{CL}_{acqua})$;
10 $\mathcal{E}$ = first γ mappings of $\mathcal{CL}_{agg}$;
11 **foreach** $\mu^{\mathcal{R}} \in \mathcal{E}$ **do**
12   μ^{S} = ServiceOp.next(JoinVars($\mu^{\mathcal{R}}$));
13   replace $\mu^{\mathcal{R}}$ with μ^{S} in $\mathcal{R}$;
14 **end**

---

mappings from "Expired.agg" list for updating, and if there is any remaining budget, it selects mappings from "Not Expired.agg" list.

Algorithm 5.2 shows the pseudo-code of the **WBM.F*** policy. For each mapping in the candidate set $\mathcal{C}$, the remaining life time, the renewed best before time, and the total score according to **WSJ-WBM** policy are computed (Lines 1–5), then, the "Expired" (Exp) and "Not Expired" ($NExp$) sets based on **WSJ-WBM** are computed (Lines 6–7). The scores of mappings are used to generate the "Expired" ($ExpL$) and "Not Expired" ($NExpL$) ranked lists (Lines 8–9).

In the next step, for each mapping in the "Expired" set (Exp), it computes the Filtering Distance as the absolute difference of the value $?x$ of mapping $\mu^{\mathcal{R}}$ and the Filtering Threshold $\mathcal{FT}$ in the query (Lines 10–12). The Filtering Distance is also computed for each mapping in the "Not Expired" set ($NExp$) (Lines 13–15). Then, it orders two sets based on the Filtering Distance (Lines 16–17) and generates the ranked lists $ExpL_f$, and $NExpL_f$. Given parameter α, and lists of mappings, the function **AggregateRanks** generates two aggregated ranked lists: "Expired.agg" ($ExpL_{agg}$) and "Not Expired.agg" ($NExpL_{agg}$) (Lines 18–19).

The set of elected mappings $\mathcal{E}$ is created by getting the top γ ones from $ExpL_{agg}list$ (Line 20). If there exists any remaining refresh budget, it gets the top mappings from $NExpL_{agg}$ list (Lines 21–24). Finally, the local replica $\mathcal{R}$ is maintained by invoking the **SERVICE** operator and querying the remote SPARQL endpoint to get fresh mappings and replace them in $\mathcal{R}$ (Lines 25–28).

5.5 Experiments

This section reports on the results of the experiments that evaluate the proposed policies. Section 5.5.1 formulates the research hypotheses. Section 5.5.2 introduces the experimental setting made of synthetic and real datasets. Section 5.5.3 provides

---

Algorithm 5.2: The pseudo-code of the WBM.F*

---

1 **foreach** $\mu^{\mathcal{R}} \in \mathcal{C}$ **do**
2 $\quad$ compute the remaining life time of $\mu^{\mathcal{R}}$;
3 $\quad$ compute the renewed best before time of $\mu^{\mathcal{R}}$;
4 $\quad$ compute the score of $\mu^{\mathcal{R}}$;
5 **end**
6 Exp = possible expired mapping of $\mathcal{C}$;
7 $NExp = \mathcal{C} - ExP$;
8 $ExpL_{wbm}$ = order Exp w.r.t. the scores;
9 $NExpL_{wbm}$ = order $NExp$ w.r.t. the scores;
10 **foreach** $\mu^{\mathcal{R}} \in Exp$ **do**
11 $\quad$ $FD(\mu^{\mathcal{R}}) = |\mu^{\mathcal{R}}(?x) - \mathcal{FT}|$;
12 **end**
13 **foreach** $\mu^{\mathcal{R}} \in NExp$ **do**
14 $\quad$ $FD(\mu^{\mathcal{R}}) = |\mu^{\mathcal{R}}(?x) - \mathcal{FT}|$;
15 **end**
16 $ExpL_f$ = order Exp w.r.t. the value of FD;
17 $NExpL_f$ = order $NExp$ w.r.t. the value of FD;
18 $ExpL_{agg} = AggregateRanks(\alpha, ExpL_f, ExpL_{wbm})$;
19 $NExpL_{agg} = AggregateRanks(\alpha, NExpL_f, NExpL_{wbm})$;
20 $\mathcal{E}$ = first γ mappings of $ExpL_{agg}$;
21 **if** $\gamma > sizeOf(ExpL_{agg})$ **then**
22 $\quad$ $\mathcal{E}'$ = first $(\gamma - sizeOf(\mathcal{E}))$ mappings of $NExpL_{agg}$;
23 $\quad$ $\mathcal{E} = \mathcal{E}$ Union $\mathcal{E}'$;
24 **end**
25 **foreach** $\mu^{\mathcal{R}} \in \mathcal{E}$ **do**
26 $\quad$ μ^S = ServiceOp.next(JoinVars($\mu^{\mathcal{R}}$));
27 $\quad$ replace $\mu^{\mathcal{R}}$ with μ^S in $\mathcal{R}$;
28 **end**

---

empirical evidence that relaxing the ACQUA.F assumption is hard, i.e., it is hard to determine a priori the band to focus on. Sections 5.5.4 and 5.5.5 report on the evaluation of the methods w.r.t. the research hypotheses and discusses the gathered practical insights.

5.5.1 Research Hypotheses

The space of evaluation has got five dimensions:

1. the proposed policies LRU.F⁺, WBM.F⁺ and WBM.F*;
2. the parameter α that allows controlling how the rank aggregation combines ACQUA and Filter Update policies in LRU.F⁺, WBM.F⁺, and WBM.F*;
3. the policies that have to compare with, i.e., Filter Update Policy, LRU.F, and WBM.F;

4. the selectivity of the filtering condition; and
5. the refresh budget available to the policies.

Notably, the parameter α and the selectivity of the filtering condition are real numbers, in the evaluation, the tests are limited to six values of α (0.167, 0.2, 0.333, 0.5, 0.0667 and 0.833). For realistic datasets, 5 values of α are selected from the range [0..1], and for synthetic data, the value of α is set to 0.2. Ten values of the selectivity (10%, 20%, ..., 90%, and 75%) is considered in the experiments. The refresh budget is, instead, an integer and as in ACQUA. Values between 1 and 7 are used, where 7 is the only value that theoretically allows refreshing all stale elements in the chosen experimental setting.

This study, in order to explore this vast space, first fixes the budget to a value, that is not enough to eliminate all stale data, and, then, tests the following two hypotheses:

Hp.3.1   For every selectivity LRU.F$^+$, WBM.F$^+$ and WBM.F* have the same accuracy of the corresponding ACQUA.F policy, but they do not require to determine a priori the band.

Hp.3.2   For every selectivity LRU.F$^+$, WBM.F$^+$ and WBM.F* are not sensitive to α, i.e., the parameter α that controls the rank aggregation can be set in a wide range of values without a significant impact on the accuracy.

In a second stage of the evaluation, the study fixes the selectivity, and tests two more hypotheses:

Hp.3.3   For every budget LRU.F$^+$, WBM.F$^+$ and WBM.F* have the same accuracy of the corresponding ACQUA.F policy.

Hp.3.4   For every budget LRU.F$^+$, WBM.F$^+$ and WBM.F* are not sensitive to α.

It is worth to note that it is not expected that LRU.F$^+$, WBM.F$^+$, and WBM.F* outperform the corresponding ACQUA.F policy, because rank aggregation can only consider the opinions of the policies it combines. In the best case, LRU.F$^+$, WBM.F$^+$ and WBM.F* can show the same accuracy of ACQUA.F. The important point is that they no longer rely on the ACQUA.F assumption that fixing the band around $\mathcal{FJ}$ is easy.

In this work, the problem is addressed if LRU.F$^+$, WBM.F$^+$, and WBM.F* are not sensitive to the parameter α, otherwise, the problem was just moved without having been solved.

5.5.2 Experimental Setting

As an experimental environment, an Intel i7 @ 1.7 GHz with 8 GB memory and a SSD disk is used. The operating system is Mac OS X 10.12.3 and Java 1.8.0.91 is installed on the machine. The experiments are done by extending the experimental setting of ACQUA.F proposed in Chap. 4.

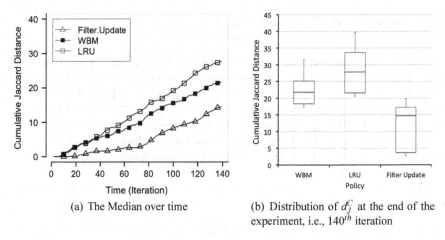

(a) The Median over time

(b) Distribution of d_J^C at the end of the experiment, i.e., 140th iteration

Fig. 5.1 The two different viewpoints that can be used to illustrate the evaluation

The experimental datasets are composed of streaming and distributed data. The streaming data is a collection of tweets from 400 verified users for three hours. The distributed data consists of the number of followers per user collected every minute.

As mentioned in Chap. 4, to control the selectivity of the filtering condition, a set of random transformations of the distributed data for a set of specified percentages are designed. To reduce the risk of bias in creating those realistic test datasets, 10 different datasets are generated for each percentage of the selectivity that are denoted with DSx% the 10 datasets where x is the selectivity.

In addition to DS10%, DS20%, ... DS90%, this study also uses six synthetic test cases namely DEC40%, DEC70%, INC40%, INC70%, MIX40% and MIX70%. The percentage refers as above to the selectivity, while INC, DEC, and MIX refer to how the number of followers of each user evolves over time. In DEC, the number of followers decreases. In INC, it always increases. In MIX, it randomly increases and decreases. In order to reduce the risk of introducing biases, each synthetic test case contains 10 different datasets.

As a test query, this study uses the one presented in Sect. 4.1 and runs 140 iterations of the query evaluation. Notably, since the query has to be evaluated once per minute the time passing is simulated for 140 min.

In order to investigate the hypotheses, this study uses the metric introduced in Sect. 2.6. As in Chap. 4, an Oracle is organized that certainly provides correct answers $Ans(Oracle_i)$ at each iteration i. These answers are compared with the possibly erroneous ones of the query $Ans(Q_i)$, and the cumulative Jaccard distance at the kth iteration for all iterations of query evaluation is computed. The lower are the value of cumulative Jaccard distance the better performances of the query evaluation.

As stated in Chap. 4, there are two points of view to show the results of the investigation of those hypotheses (Fig. 5.1). The first point of view takes a time-series perspective and it allows comparing the accuracy of the various policies through the

Table 5.1 Summary of FDT value in case the policy has minimum Cumulative Jaccard Distance

	INC40%	INC70%	DEC40%	DEC70%	MIX40%	MIX70%
LRU.F	156	1000	250	625	500	500
WBM.F	500	1000	375	375	109	93

time for every evaluation. For instance, Fig. 5.1a shows the medians of cumulative Jaccard distance over time for **WSJ-WBM**, **WSJ-LRU** and **Filter Update Policy** when tested with DS75% and a refresh budget of 3. The plot shows that each policy has got constant behavior over time; for example, the **Filter Update Policy** is the best policy for each iteration.

The second point of view (Fig. 5.1b) focuses on the distribution of the cumulative Jaccard distance at the end of the experiment (at the $140th$ iteration). It uses a box-plot [14] to highlight the median and the four quartiles of the accuracy obtained running the experiments against the 10 datasets in the DS75% test case. The box-plot shows that for the first three quartiles the Filter Update policy is more accurate than all others; only the first quartile of **WSJ-WBM** has a comparable accuracy. This section focuses only on the second point of view to show the results of the experiments.

In order to evaluate hypotheses Hp.3.2 and Hp.3.4, a statistic t-test is set up for different values of alpha to determine if the two sets of results are significantly different from each other or not. Independent samples t-test is applied, which compares the means for two groups of data ($\mu 1$, and $\mu 2$). The null hypothesis for the independent samples t-test is $\mu 1 = \mu 2$. In these tests, the confidence interval is set to 0.95. The p-value obtained from the t-test is a number between 0 and 1 that shows the strength of the evidence against the null hypothesis. If the p-value is greater than 0.1, the data are consistent with the null hypothesis, therefore the two sets of data are not significantly different from each other. Moreover, if the p-value is small enough (p-value < 0.1), then there exists enough evidence (weak, moderate, strong, very strong) to reject the null hypothesis, so the differences between the two sets of data are statistically significant.

5.5.3 Relaxing ACQUA.F Assumption

This experiment provides empirical evidence that relaxing the **ACQUA.F** assumption is hard. **ACQUA.F** assumes that it is simple to determine a priori the Filter Distance Threshold (FDT), i.e., the band around the Filtering Threshold to focus on.

To check if relaxing this assumption is easy, **LRU.F** and **WBM.F** policies are tested with the DEC40%, DEC70%, INC40%, INC70%, MIX40%, and MIX70% test cases. Each test case runs several times with different FDT values. The refresh budget γ is equal to 3.

Fig. 5.2 Result of
experiment 1 that runs rank
aggregation policies over
synthetic datasets to compare
them with existing policies
for different selectivities

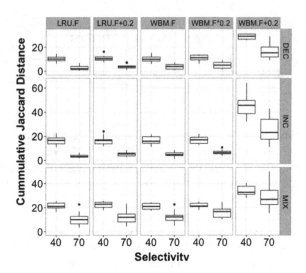

Table 5.1 summarizes for each policy and test case the value of FDT in which the
cumulative Jaccard distance is minimal. The results show that relaxing the assumption
of knowing FDT is hard. The **ACQUA.F** policies are sensitive to FDT and fixing
a single FDT is not straightforward.

5.5.4   Experiment 1—Sensitivity to the Filter Selectivity

This experiment tests hypothesis Hp.3.1 and Hp.3.2 by checking the sensitivity to the
filter selectivity for the proposed policies considering different values of α. Keeping
the refresh budget γ equal to 3, experiments run on both synthetic and real test cases.
As explained in Sect. 5.5.2, the results are presented using box-plots that capture the
distribution of the cumulative Jaccard distance at the end of the experiment. The
results are obtained using the 10 different datasets that are contained in each test
case.

Figure 5.2 shows the results when using synthetic datasets. Each column shows the
results related to one policy and for two levels of selectivity (40 and 70%). Each row
shows the experiments run using the same test case (DEC, INC, and MIX). To show
the value of α used in the rank aggregation policies, the notation $< policy > < \alpha >$
is used. For example, LRU.F$^+$0.2 in the second column refers to LRU.F$^+$ policy
when using $\alpha = 0.2$.

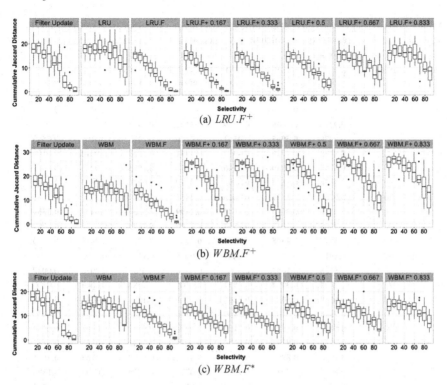

Fig. 5.3 Result of experiment 1 that runs rank aggregation policies over real datasets to compare them with existing policies for different selectivities

The results of the first two columns show that LRU.F$^+$ with α equal to 0.2 is comparable to LRU.F for all synthetic test cases (i.e., DEC, INC, and MIX). The results of columns three to five show that WBM.F* with $\alpha = 0.2$ is comparable to WBM.F, while WBM.F$^+$ with $\alpha = 0.2$ is worst than WBM.F.

Overall the results allow saying that, w.r.t. the synthetic test cases, Hp.3.1 is verified only by LRU.F$^+$ and WBM.F*. In other words, the best approach for combining WSJ-WBM policy and Filter Update Policy is considering the "Expired" and "Not Expired" lists of WSJ-WBM in rank aggregation algorithm (see Algorithm 5.2).

Figure 5.3 shows the results obtained for the test cases DS10%, DS20%, ..., and DS90%. Each column shows the results related to one policy for different selectivities. The first three columns show the results of Filter Update Policy, ACQUA and ACQUA.F policies, respectively. Columns four to eight show the results for the proposed rank aggregation policies for five different values of α.

Figure 5.3a compares the proposed LRU.F$^+$ policy with Filter Update Policy, WSJ-LRU, and LRU.F. Independently from the selectivity, LRU.F$^+$ with $\alpha = 0.167$ is as accurate as LRU.F and remains better than Filter Update Policy and WSJ-LRU. This verifies Hp.3.1 w.r.t. LRU.F$^+$ on real data.

Table 5.2 Summary of statistic tests for different policies and selectivities, bold faces cells high light values of α and selectivity for which the proposed policies are not sensitive to α

Policy	Selectivity	p-value						
		$\alpha = 0.167$ versus $\alpha = 0.333$	$\alpha = 0.167$ versus $\alpha = 0.5$	$\alpha = 0.333$ versus $\alpha = 0.5$	$\alpha = 0.167$ versus $\alpha = 0.667$	$\alpha = 0.5$ versus $\alpha = 0.667$	$\alpha = 0.167$ versus $\alpha = 0.833$	$\alpha = 0.5$ versus $\alpha = 0.833$
WBM.F*	90	**0.8769**	**0.7076**	**0.8297**	**0.1796**	**0.3226**	0.0148	0.0370
	80	**0.9273**	**0.9949**	**0.9249**	**0.2122**	**0.1593**	0.0062	0.0042
	75	**0.9028**	**0.3965**	**0.4561**	**0.2240**	**0.7181**	0.0109	0.0553
	70	**0.8283**	**0.3310**	**0.4354**	**0.1441**	**0.5960**	0.0010	0.0100
	60	**0.6705**	**0.3980**	**0.6746**	0.0471	**0.1815**	0.0013	0.0087
	50	**0.6354**	**0.1583**	**0.3392**	0.0719	**0.6772**	0.0113	**0.2242**
	40	**0.6751**	**0.5527**	**0.8423**	**0.2851**	**0.5894**	0.0267	**0.1047**
	30	**0.8818**	**0.3576**	**0.2741**	**0.1499**	**0.5535**	0.0435	**0.2359**
	20	**0.4605**	**0.1615**	**0.4864**	**0.2120**	**0.7179**	0.0184	**0.3007**
	10	**0.5897**	**0.4308**	**0.8114**	**0.2384**	**0.6744**	**0.2471**	**0.6730**
WBM.F+	90	**0.1914**	0.0654	**0.4449**	0.0046	**0.1120**	0.0009	0.0140
	80	**0.5208**	**0.1697**	**0.3287**	0.0030	0.0137	0.0010	0.0042
	75	**0.7424**	**0.3224**	**0.4937**	0.0168	0.0684	0.0043	0.0200
	70	**0.6559**	**0.5086**	**0.9201**	0.0561	**0.1239**	0.0045	0.0098
	60	**0.9548**	**0.3406**	**0.3060**	0.0766	**0.3447**	0.0053	0.0436
	50	**0.8670**	**0.5925**	**0.7132**	0.0222	0.0772	0.0091	0.0284
	40	**0.9913**	**0.8314**	**0.8370**	**0.1174**	**0.1541**	0.0853	**0.1127**
	30	**0.9934**	**0.8899**	**0.8834**	**0.3423**	**0.4378**	**0.2163**	**0.2885**
	20	**0.9705**	**0.7033**	**0.7319**	0.0774	**0.1499**	0.0993	**0.1808**
	10	**0.9293**	**0.8666**	**0.9363**	**0.6619**	**0.7781**	**0.5695**	**0.6752**
LRU.F+	90	0.0347	0.0014	0.0186	0.0004	0.0078	0.0037	0.0271
	80	0.0891	0.0081	0.0797	0.0004	0.0121	0.0000	0.0000
	75	**0.1105**	0.0003	0.0256	0.0003	0.0165	0.0000	0.0009
	70	0.0561	0.0001	0.0223	0.0000	0.0006	0.0001	0.0053
	60	**0.1200**	0.0183	**0.2853**	0.0042	**0.1389**	0.0000	0.0001
	50	**0.2030**	0.0843	**0.6765**	0.0031	**0.1712**	0.0001	0.0050
	40	**0.6984**	**0.4154**	**0.6800**	0.0442	**0.2036**	0.0005	0.0040
	30	**0.7426**	**0.1951**	**0.3570**	0.0313	**0.3222**	0.0008	0.0096
	20	**0.8711**	**0.7973**	**0.9246**	**0.5946**	**0.7740**	0.0994	**0.1605**
	10	**0.9060**	**0.9170**	**0.9857**	**0.7923**	**0.7128**	**0.5670**	**0.4994**

The experiments on the synthetic and the real test cases show that LRU.F$^+$ can have practical value for low selectivities, because it works for a wide range of values of α (0.167 − 0.5).

Figure 5.3b allows comparing the proposed WBM.F$^+$ with different values of α with Filter Update Policy, WSJ-WBM, and WBM.F. The box-plots show that WBM.F$^+$ is less accurate than WBM.F, and Filter Update Policy. Therefore, Hp.3.1 is not verified for WBM.F$^+$.

Fig. 5.4 Result of experiment 2 that runs rank aggregation policies over synthetic datasets to compare them with existing policies for different refresh budgets

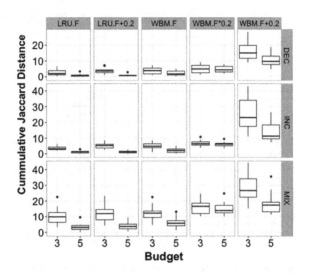

The experiments show that merging the two lists of "Expired" and "not Expired" mappings in the WSJ-WBM algorithm can badly affect the result. WBM.F$^+$ is of no practical usage.

Figure 5.3c allows comparing the proposed WBM.F* with different values of α with Filter, WBM, and WBM.F. WBM.F* is as accurate as WBM.F for selectivities smaller than 60%. Accordingly, Hp.3.1 w.r.t. WBM.F* is only partially verified for low selectivities.

Table 5.2 shows the result of the t-tests for different policies and different values of alpha to verify Hp.3.2. For policies WBM.F$^+$, and WBM.F*, all the p-values in columns three to five are greater than 0.1 (bold numbers): for alpha values equal to 0.167, 0.333, and 0.5, there is not enough evidence to show significant differences between policies. The rest of the columns have p-values smaller than 0.1, which give enough evidence to reject the null hypothesis, so the differences between the policies are statistically significant. For LRU.F$^+$ policy, only for low selectivity (<50), all the p-values in the columns three to five are greater than 0.1.

So, policies WBM.F$^+$, and WBM.F*, with alpha values equal to 0.167, 0.333, and 0.5 are similar to each other, which verifies Hp.3.2 w.r.t. WBM.F$^+$, and WBM.F* on real data. Hp.3.2 is also verified w.r.t. LRU.F$^+$ on real data for low selectivities.

From a practical perspective, it is worth observing that WBM.F* with $\alpha = 0.167$ is: (*i*) always better than WSJ-WBM (i.e., the best policy in ACQUA) and (*ii*) better than LRU.F$^+$ for low selectivity. Having to choose a policy, LRU.F$^+$ is the one that on average gives the best accuracy, but having the possibility to estimate the selectivity at run time, it would be better to use WBM.F* for low selectivities (<60%) and LRU.F$^+$ for high selectivities ($\geq$60%).

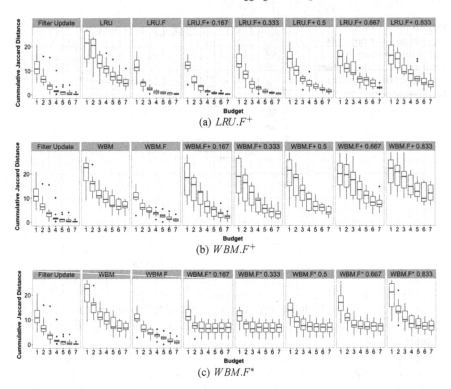

Fig. 5.5 Result of experiment 2 that runs rank aggregation policies over real datasets to compare them with existing policies for different refresh budgets

5.5.5 Experiment 2—Sensitivity to the Refresh Budget

This experiment tests Hp.3.3 and Hp.3.4 by investigating the sensitivity to the refresh budget γ for the proposed policies and for different values of α.

These experiments run over a subset of the test cases introduced in Sect. 5.5.2. For synthetic data, the experiments run for refresh budget 3 and 5 over the DEC70%, INC70%, and MIX70% test cases. For the real data, the refresh budget varies from to 1 to 7 and the experiments run over DS75% test case.

Selectivity is fixed to 70% for the synthetic data and 75% for the real data, because, according to the results reported in Sect. 5.5.4, this is the smallest value of selectivity for which LRU.F$^+$, WBM.F$^+$, and WBM.F* have comparable accuracy.

Figure 5.4 shows the results obtained using synthetic test cases. The first two columns allow asserting that the result of LRU.F and LRU.F$^+$ with $\alpha = 0.2$ are comparable. The columns three to five show that WBM.F* and WBM.F$^+$ with $\alpha = 0.2$ are worse than WBM.F. In particular, WBM.F$^+$ is the worst for both budget, whereas WBM.F* seems unable to use additional budget (i.e., the accuracy with

Table 5.3 Summary of statistic tests for different policies and refresh budgets, bold faces cells high light values of α and refresh budget for which the proposed policies are not sensitive to α

Policy	Budget	p-value						
		$\alpha = 0.167$ versus $\alpha = 0.333$	$\alpha = 0.167$ versus $\alpha = 0.5$	$\alpha = 0.333$ versus $\alpha = 0.5$	$\alpha = 0.167$ versus $\alpha = 0.667$	$\alpha = 0.5$ versus $\alpha = 0.667$	$\alpha = 0.167$ versus $\alpha = 0.833$	$\alpha = 0.5$ versus $\alpha = 0.833$
$WBM.F^*$	1	**0.9376**	**0.2290**	**0.1527**	0.0202	**0.1351**	0.0020	0.0142
	2	**0.8011**	**0.2353**	**0.2935**	0.0080	**0.1916**	0.0007	0.0097
	3	**0.9028**	**0.3965**	**0.4561**	0.2240	**0.7181**	0.0109	0.0553
	4	**0.8593**	**0.5623**	**0.6860**	0.2675	**0.6308**	0.0946	0.2677
	5	**0.9641**	**0.7472**	**0.7823**	0.3353	**0.5073**	**0.2314**	**0.3700**
	6	**0.9777**	**0.8910**	**0.9136**	**0.6506**	**0.7446**	**0.4634**	**0.5397**
	7	**0.9691**	**0.9020**	**0.9297**	**0.6982**	**0.7868**	**0.5442**	**0.6176**
$WBM.F^+$	1	**0.8276**	**0.6150**	**0.7749**	**0.1450**	**0.3128**	0.0891	**0.2106**
	2	**0.8248**	**0.4232**	**0.5764**	0.0508	**0.2252**	0.0313	**0.1373**
	3	**0.7424**	**0.3224**	**0.4937**	0.0168	0.0684	0.0043	0.0200
	4	**0.5610**	**0.2676**	**0.5621**	0.0044	0.0340	0.0003	0.0022
	5	**0.4062**	**0.1115**	**0.4119**	0.0005	0.0101	0.0000	0.0004
	6	**0.3752**	0.0128	**0.1364**	0.0005	0.0316	0.0001	0.0019
	7	**0.1443**	0.0087	**0.3610**	0.0003	0.0138	0.0000	0.0001
$LRU.F^+$	1	**0.2801**	**0.1853**	**0.7454**	0.0505	**0.5070**	0.0435	**0.4231**
	2	0.0338	0.0039	**0.2160**	0.0002	**0.4903**	0.0000	0.0181
	3	0.1105	0.0003	0.0256	0.0003	0.0165	0.0000	0.0009
	4	0.0095	0.0015	0.0270	0.0000	0.0858	0.0000	0.0108
	5	0.0307	0.0000	0.0003	0.0001	0.0217	0.0002	0.0051
	6	**0.3080**	0.0006	0.0016	0.0000	0.0014	0.0002	0.0032
	7	**0.1366**	0.0025	0.0093	0.0001	0.0098	0.0002	0.0022

budget 5 is similar to the accuracy with budget 3). Therefore, Hp.3.3 is verified for LRU.F$^+$, but not for WBM.F$^+$ and WBM.F*.

This observation provides additional insight on WBM.F$^+$ and WBM.F*. In discussing Hp.3.1, Sect. 5.5.4 noted that WBM.F* is more accurate than WBM.F$^+$ for budget 3, but giving more budget to WBM.F* does not turn in more accurate results.

Real data (see Fig. 5.5) confirms the insights gathered using synthetic data: LRU.F$^+$ is comparable with LRU.F (Fig. 5.5a), while WBM.F$^+$ and WBM.F* are worse than WBM.F (Fig. 5.5b and c). WBM.F* is not able to use all the budget when it is greater than 3. On the contrary, WBM.F$^+$, given a high budget, becomes comparable to WBM.F. Therefore, Hp.3.3 is verified for LRU.F$^+$, partially verified for WBM.F$^+$ for the budget greater than 5, and not verified for WBM.F*.

From a practical perspective, this analysis confirms that, if one has to choose a policy, LRU.F$^+$ is on average the best one. WBM.F* is a perfect solution only when the available budget is very low.

Table 5.3 shows the result of the t-tests for different policies and refresh budgets to verify Hp.3.4. For policies WBM.F$^+$, and WBM.F* almost all the p-values in columns three to five are greater than 0.1 (bold numbers), so, for alpha values equal to 0.167, 0.333, and 0.5, there is no significant differences between policies, and Hp.3.4 is verified. But for policy LRU.F$^+$, most of the p-values are less than 0.1, so the differences between policies are statistically significant. Therefore, LRU.F$^+$ is sensitive to α for different refresh budgets.

5.6   Outlook

This chapter further investigates the ACQUA.F approach by removing the assumption that it is possible to determine a priori the band to focus on. The proposed new policies use rank aggregation. Those new policies let each ACQUA.F policy express its opinion by ranking data items according to its own criterion and, then, aggregate those ranks to take fairly into account all opinions.

To study the research question, four hypotheses were formulated. Hypotheses Hp.3.1 and Hp.3.3 test if the proposed policies have the same accuracy of the corresponding ACQUA.F policies, without determining a priori the band to focus on. Hypotheses Hp.3.2, and Hp.3.4 test if the proposed policies are sensitive to α. The results are reported in Table 5.4.

The results of experiment 1 (about Hypotheses Hp.3.1, and Hp.3.2) show that LRU.F$^+$ policy has the same accuracy of the LRU.F policy for every selectivity, and WBM.F* policy is comparable to WBM.F policy for low selectivity. They also show that WBM.F$^+$ and WBM.F* policies are little sensitive to α and $\alpha \in [0.167, 0.5]$ is acceptable for every selectivity.

The results of experiment 2 (about Hypotheses Hp.3.3, and Hp.3.4) show that LRU.F$^+$ policy has the same accuracy of the LRU.F policy for every budget, and WBM.F$^+$ policy is comparable to WBM.F policy for the high value of budget. They also show that WBM.F* is not able to use all the budget, and even increasing the budget, the error does not go below a given minimum. Moreover, WBM.F$^+$, and WBM.F* policies are not sensitive to α.

The following section focuses on the class of top-k queries. The proposed policies in ACQUA, ACQUA.F are not tailored for top-k queries. Moreover, the processing of top-k queries over streaming data has its own challenges, which are not addressed in previous work.

Table 5.4 Summary of the verification of the hypotheses w.r.t. LRU.F$^+$, WBM.F$^+$, and WBM.F*

	Measuring	Varying	$LRU.F^+$	$WBM.F^+$	$WBM.F^*$
Hp.3.1	Accuracy	Selectivity	✓		<60%
Hp.3.2	Sensitivity to α	selectivity, α	<50%	✓	✓
Hp.3.3	Accuracy	budget	✓	>5	
Hp.3.4	Sensitivity to α	budget, α		✓	✓

References

1. Bansal N, Blum A, Chawla S (2004) Correlation clustering. Mach Learn 56(1–3):89–113
2. Berry G (1989) Real time programming: special purpose or general purpose languages. PhD thesis, INRIA
3. de Borda JC (1781) Mémoire sur les élections au scrutin
4. De Condorcet N et al (2014) Essai sur l'application de l'analyse à la probabilité des décisions rendues à la pluralité des voix. Cambridge University Press
5. Della Valle E, Dell'Aglio D, Margara A (2016) Taming velocity and variety simultaneously in big data with stream reasoning: tutorial. In: Proceedings of the 10th ACM international conference on distributed and event-based systems. ACM, pp 394–401
6. Diaconis P (1988) Group representations in probability and statistics. Lecture Notes-Monograph Series 11:i–192
7. Diaconis P, Graham RL (1977) Spearman's footrule as a measure of disarray. J R Statistical Soc Ser B (Methodol), pp 262–268
8. Dwork C, Kumar R, Naor M, Sivakumar D (2001) Rank aggregation methods for the web. In: WWW. ACM, pp 613–622
9. Fagin R, Kumar R, Mahdian M, Sivakumar D, Vee E (2006) Comparing partial rankings. SIAM J Discret. Math 20(3):628–648
10. Fagin R, Kumar R, Sivakumar D (2003) Comparing top k lists. SIAM J Discret Math 17(1):134–160
11. Fagin R, Lotem A, Naor M (2003) Optimal aggregation algorithms for middleware. J Comput Syst Sci 66(4):614–656
12. Kemeny JG (1959) Mathematics without numbers. Daedalus 88(4):577–591
13. Kemeny JG (1972) Mathematical models in the social sciences. Technical report
14. McGill R et al (1978) Variations of box plots. Am Statist 32(1):12–16

Chapter 6
Handling Top-k Queries

Abstract This chapter focuses on continuously finding the most relevant (shortly, top-k) answer of a query that joins streaming and distributed data. Remaining reactive can be challenging, when accessing the distributed data has high latency and is rate limited. As illustrated in the previous chapters reactiveness of the system can be guaranteed by using a local replica of the distributed data. This chapter introduces the proposed framework for Top-k query evaluation. To overcome the problem of stale data in the local replica, two maintenance policies are proposed: one maximizes the relevancy of the result, while the other one maximizes the accuracy of the top-k result. Empirical evidence shows that the proposed policies produce more relevant and accurate results than state-of-the-art ones.

Keywords Continuous query answering · Top-k query answering · Streaming data · Evolving distributed data · RDF data · RSP engine

6.1 Introduction

Many Web applications require to combine dynamic data streams with data distributed over the Web to continuously answer queries. Consider the following examples. In social content marketing, advertisement agencies may want to continuously detect influential Social Network users, when they are mentioned in micro-posts across Social Networks, in order to ask them to endorse their commercials. As mentioned in Chap. 1, the number of followers may change in seconds, and the result of the query should be returned in a minute, otherwise, the competitors may reach the influencer sooner.

Here is another example. Finding a parking lot could be difficult especially in big cities or crowded places such as city centers. In Smart Cities domain, users may want to predict the availability of parking lots based on the information of parking spaces, data detected through smartphone, sensors, or cameras and descriptions of points of interest and events [1]. Drivers may benefit from an application[1] that shows

---

[1]Easypark activated a similar service in Stockholm, but it still relays on the centralized system.

© The Author(s), under exclusive license to Springer Nature Switzerland AG 2020   75
S. Zahmatkesh and E. Della Valle, *Relevant Query Answering over Streaming and Distributed Data*, PoliMI SpringerBriefs,
https://doi.org/10.1007/978-3-030-38339-8_6

the places around them where there is a high probability of finding free parking. The application needs to keep static data such as the map of the city, and the positions of the parking lots. The positions of the cars continuously change (every second) and can be seen as a stream. On the contrary, the data related to the free parking lots slowly evolves (changes every minute) and can be seen as a part of the distributed data. The application detects areas with high probability of finding free parking lots by maximizing the number of free parking lots in the area and minimizing the number of cars it has already rooted to that specific area. It is possible to formulate the solution to this problem as a continuous top-k query such as:

Return the areas (around the car that calls the service) where there are the largest number of free parking lots and the smallest number of cars looking for parking in the last 10 min.

As in the Social Media scenario, answering this query in a reactive manner is challenging due to high latency and rate limits in accessing the distributed data over the Web.

However, one may wonder if it is really necessary to perform the entire join to answer those two information needs introduced above. They clearly focus only on the top results. Indeed, the state of the art includes partial solutions to this problem. The database community studied *continuous top-k queries over the data streams* [19] that handles massive data streams focusing only on the top-k answer but ignores the join with evolving distributed data.

The Semantic Web community showed that RDF Stream Processing (RSP) engines provide an adequate framework for continuous joining of streaming and distributed data [3]. In this setting, distributed data is usually accessible by using SPARQL query over SPARQL endpoints. In order to access remote services, the query has to use federated SPARQL syntax [16] which is supported by different RSP query languages (e.g. RSP-QL [5]).

As stated in Chap. 3, ACQUA studies *approximate continuous joining of RDF streams and dynamic linked data sets* [2] which is reactive by design but it is not optimized for top-k queries.

For instance, Listing 6.1 shows a simplified version of the first example above, which is formulated as a continuous RSP-QL top-k query using the syntax proposed in [4]. Top-k Queries get a user-specified scoring function, and provide only the top-k answer with the highest score based on the scoring function.

```
1   REGISTER  STREAM  : TopkUsersToContact  AS
2   SELECT  ?user
3            F(?mentionCount ,? followerCount)  as  ?score
4   FROM  NAMED  WINDOW  :W  ON  :S  [RANGE  9m  STEP  3m]
5   WHERE{
6    WINDOW  :W  {?user  :hasMentions  ?mentionCount}
7    SERVICE  :BKG  {?user  :hasFollowers  ?followerCount}
8   }
9   ORDER  BY  DESC  (?score)
10  LIMIT  1
```

Listing 6.1 A top-k continues RSP-QL query that joins streaming and distributed data

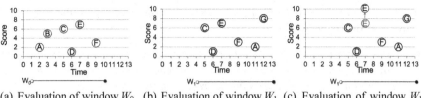

(a) Evaluation of window W_0 (b) Evaluation of window W_1 (c) Evaluation of window W_1 considering changing

Fig. 6.1 The example that shows the objects in the top-k result after join clause evaluation of windows W_0, and W_1

At each query evaluation, the WHERE clause at Lines 5–8 is matched against the data in a window :W opens on the data stream :S, on which the mentions of each user flow, and in the remote SPARQL service :BKG, which contains the number of followers for each user. Function F computes the score of each user as the normalized sum of his/her mentions (?mentionCount) and his/her number of followers (?followerCount). The users are ordered by their scores, and the number of results is limited to 1.

Figure 6.1a, b show a portion of a stream between time 0 and 13. The X axis shows the arriving time on the stream of the number of mentions of a certain user to the system, while the Y axis shows the score of the user computed after evaluating the join clause with the number of followers fetched from the distributed data. For the sake of clarity, each point is labeled in the Cartesian space with the ID of the user it refers to. This stream is observed through a window that has a length equal to 9 min and slides every 3 min. In particular, Fig. 6.1a shows the content of window W_0 that opens at 1 and close at 10 (excluded). Figure 6.1b shows the next window W_1 after the sliding of 3 min. Each circle indicates the score of a user after the evaluation of the JOIN clause, but before the evaluation of the ORDER and LIMIT clauses.

During window W_0 users A, B, C, D, E, and F come to the system (Fig. 6.1a). When W_0 expires, users A and B go out of the result. Before the end of window W_1, the new users G, and H appear (Fig. 6.1b). Evaluating the query in Listing 6.1 gives us user E as the top-1 result for window W_0 and user G as the top-1 result for window W_1.

However, changes in the number of followers of a user in the distributed data can change the score of a user between subsequent query evaluations, and this can affect the result. For example, in Fig. 6.1c, between the evaluation time of windows W_0, and W_1, the score of user E changes from 7 to 10 (due to the changes in the number of followers in the distributed data). Considering the new score of user E in the evaluation of window W_1, the top-1 result is no longer user G, but it changes to user E.

As mentioned above, while RSP-QL allows encoding top-k queries, state-of-the-art RSP engines are not optimized for such a type of queries and they would recompute the result from scratch as explained in [14, 15]. This put them at risk of losing reactiveness. In order to handle this situation, in this chapter, the following research question is investigated:

RQ.4 *How to optimize continuous, if needed approximate, top-k queries that join streaming and distributed data, which may change between two consecutive evaluations, while guaranteeing the reactiveness of the system?*

In continuous top-k query answering, it is well known that recomputing the top-k result from scratch at every evaluation is a major performance bottleneck. In 2006, Mouratidis et al. [14] were the first to solve this problem proposing an incremental query evaluation approach that uses a data structure known as k-skyband and an algorithm to precompute the future results in order to reduce the probability of recomputing the top-k results from scratch.

A few years after, in 2011, Di Yang et al. [19] completely removed this performance bottleneck designing *MinTopk* algorithm which answers a top-k query without any recomputation of top-k results from scratch. The approach memorizes only the minimal subset of the streaming data which is necessary and efficient for query evaluation and discards the rest. The authors also showed the optimality of the proposed algorithm in both CPU and memory utilization for continuous top-k monitoring. Although *MinTopk* algorithm presents an optimal solution for top-k query answering over the data stream, it did not consider joining with distributed data, aggregated score, distinct arrival of items, and changes in scoring values. Therefore, MinTopk algorithm does not work properly in such cases.

As introduced in the previous chapters, a solution to this problem can be found for RSP engines by introducing a replica of the distributed data. A *refresh budget* limits the number of remote accesses to the distributed data for updating the local replica, and guarantees by construction the reactiveness of the system. However, if the refresh budget is not enough to refresh all the data in the replica, some of the data items become stale, and the query result can contain errors. To avoid this Chaps. 3, 4, and 5 introduce designed maintenance policies that update the local replica in order to reduce the number of errors and approximate the correct result. Unfortunately, also none of those policies are optimized for top-k queries.

This chapter extends the state-of-the-art approach for top-k query evaluation [19], considering distributed data that changes during the evaluation. The contributions are highlighted in italics in the following paragraphs.

A first solution assumes that all changes are pushed from the distributed data to the engine that continuously evaluates the query. It extends the data structure proposed in [19] and introduces that Super-MTK+N list to keep the necessary and sufficient data for top-k query evaluation. The proposed data structure can handle changes in distributed data while *minimizing the memory usage*. However, MinTopk algorithm [19] assumes distinctive arrivals of data. So, to handle the changes pushed from the distributed data, it needs modification to support *indistinct arrivals* of data. Indeed, in the example, user E is already in the window when his/her number of followers changes and so does his/her score. The proposed Topk+N *algorithm* considers the changed data items as new arrivals with new scores.

This first solution works in a data center, where the entire infrastructure is under control, network latency is low, and bandwidth is large. However, it may not work on the Web, which is decentralized and where high network latency, low bandwidth, and even rate-limited access are frequently experienced. In this setting, the engine, which continuously evaluates the query, has to pull the changes form the distributed data.

Therefore, considering the architectural approach presented in Chap. 3 as a guideline, this chapter proposes the **AcquaTop** *framework*. Notably, when there is enough refresh budget to update all the stale elements in the replica, the proposed approach is exact, but when there is not, the result might have some errors.

In order to approximate as much as possible the correct answer in this extreme situation, two maintenance policies are proposed to update the replica using **AcquaTop** *algorithm*. They are specifically tailored to top-k approximated join. *Top Selection Maintenance* (**AT-TSM**) *policy*, maximize the relevance, i.e., minimizes the difference between the order of the answers in the approximate top-k result and the correct order. *Border Selection Maintenance* (**AT-BSM**) *policy*, instead, maximizes the accuracy of the top-k result, i.e., it tries to get all the top-k answers in the result, but it ignores the order.

The remainder of the chapter is organized as follows. Section 6.2 formalizes the problem. Sections 6.3, and 6.4 present the proposed solution for top-k query evaluation over streaming and dynamic distributed data. Section 6.5 discusses the experimental setting and the research hypotheses, reports on the evaluation of the proposed approach and highlights some practical insights. Section 6.6 reviews the related work regarding the contributions and, Sect. 6.7 concludes the achievements.

6.2 Problem Statement

This chapter considers top-k continuous RSP-QL queries over a data stream S and a distributed data $\mathcal{D}$ as in previous chapters. Here are the assumptions: (i) there is a 1:1 join relationship between the data items in the data stream and those in the distributed dataset; (ii) the window, opened over the stream S, slides (i.e., $\omega > \beta$); (iii) queries are evaluated when windows close, and (iv) the distributed data evolves between subsequent evaluations.

Moreover, the algebraic expression SE of this class of RSP-QL queries is defined as in Fig. 6.2a, where:

- P_S, and $P_{\mathcal{D}}$ are graph patterns,
- u_S, and $u_{\mathcal{D}}$ identify the window on the RDF stream and the remote SPARQL endpoint,
- μ_S is a solution mapping of the graph pattern *WINDOW* u_S P_S,
- $\mu_{\mathcal{D}}$ is a solution mapping of the graph pattern *SERVICE* $u_{\mathcal{D}}$ $P_{\mathcal{D}}$,
- x_S, and $x_{\mathcal{D}}$ are scoring variables in mapping μ_S and $\mu_{\mathcal{D}}$,
- $x_{\mathcal{J}}$ is a join variable in $dom(\mu_S) \cap dom(\mu_D)$, and

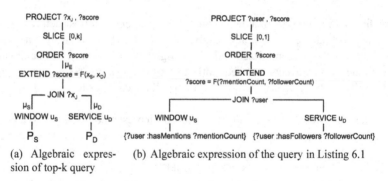

(a) Algebraic expression of top-k query

(b) Algebraic expression of the query in Listing 6.1

Fig. 6.2 Algebraic expression

- $F(x_S, x_D)$ is a monotone scoring function, which generates the score and adds it to the solution mapping by using the EXTEND operator.

For the sake of clarity, Fig. 6.2b illustrates the algebraic expression of the query in Listing 6.1. *?user :hasMentions ?mentionCount*, and *?user :hasFollowers ?followerCount* are the graph patterns respectively in the WINDOW and in the SERVICE clauses. *?mentionCount*, and *?followerCount* are the scoring variables, and *?user* is the join variable. The scoring function F gets *?mentionCount*, and *?followerCount* as inputs and generates the score for each user. The values of *?mentionCount*, and *?followerCount* can increase or decreased over time, and any linear combination of these two variables can guarantee to have a monotonic function. Once each solution mapping of the join is extended with a score, the solution mappings are ordered by their score and the top-k ones are reported as a result.

In the remainder of the chapter, considering the solution mappings Ω_E of the EXTEND graph pattern, for each solution mapping $\mu_E \in \Omega_E$: $dom(\mu_E) = dom(\mu_S) \cup dom(\mu_D) \cup \{?score\}$. Let us call *Object O(id, score)* one of such results, where the $id = \mu_E(x_J)$, and the score $O.score$ is a real number computed by the scoring function $F(\mu_E(x_S), \mu_E(x_D))$. $O.score_S$, and $O.score_D$ denote the values coming from the streaming and the dynamic distributed data, respectively, i.e., $O.score_S = \mu_E(x_S)$, and $O.score_D = \mu_E(x_D)$.

Let us, now, formalize the notion of changes in the distributed data that may occur between two consecutive evaluations of the top-k query. Assuming et' and et'' as two consecutive evaluation times (i.e. et', $et'' \in ET$, and $\nexists et''' \in ET : et' < et''' < et''$) the instantaneous graph $\overline{G}_d(et')$ in the distributed data differs from the instantaneous graphs $\overline{G}_d(et'')$.

Those changes in the values of the scoring variables of objects, which are used to compute the scores, can affect the result of the top-k query. Assuming that $O.score_{et'}$ is the score of object O at time et', and $O.score_{et''}$ is the score of object O at time et''. $O.score_{et''}$ may be different from $O.score_{et'}$ due to the changes in the value of $\mu_E(x_D)$ that comes from the distributed data.

Therefore, in the evaluation of the query at time et'', it is not possible to count on the result obtained in the previous evaluation, as the score of object O at the evaluation time et' may differ from the one at time et'' and this can give us an incorrect answer."

For instance, in the example of Fig. 6.1c, the score of object E changes from 7 to 10 between the evaluation of windows W_0, and W_1. So, the top-1 result of window W_1 is object E instead of object G.

Let denote $Ans(Q_i)$ the possibly erroneous answer of the query evaluated at time i. If, for every query evaluation, the join is recomputed and the score of each object is generated from scratch, all iterations have the correct answer. Let denote with $Ans(RQ_i)$ the correct answer for iteration i denotes as.

As stated in Sect. 2.6, for each iteration i of the query evaluation, it is possible to compute the $nDCG@k$ and $ACC@k$ comparing the query answer $Ans(Q_i)$, and the correct answer $Ans(RQ_i)$. The higher value of $nDCG@k$ and $ACC@K$ show respectively more relevant and accurate results. Let us denote with M the set of metrics $\{nDCG@k, ACC@K\}$ and define the error as follow:

$$error = 1 - M \tag{6.1}$$

So, the goal in this chapter is to approximate results, i.e., to minimize the error.

For example, assuming that the correct answer $Ans(RQ_i)$ is equal to $\{A, B, C\}$, and the query answer $Ans(Q_i)$ is equal to $\{A, D, F\}$, as mentioned in Sect. 2.6, the $nDCG@3$ is equal to 0.754 and the $precision@3$ is equal to 0.333, and the respective errors are $1 - 0.754 = 0.246$ and $1 - 0.333 = 0.667$.

6.3 Topk+N Solution

This section introduces the proposed solution to the problem of top-k query answering over streaming and distributed data in the context of RSP engines. Being reactive is the most important requirement, while the distributed data changes. Section 6.3.1 shows how the approach in [19] is extended for streaming and distributed data. Section 6.3.2 introduces the MTK+N data structure. Section 6.3.3 introduces the Super-MTK+N list. Finally, Sect. 6.3.4 explains the Topk+N algorithm, which is optimized for top-k query answering.

6.3.1 MinTopk+

As mentioned in Sect. 2.5, MinTopk [19] offers an optimal strategy to monitor top-k queries over streaming data. This first subsection reports on how to extend [19] so to handle changes in the distributed data.

In the setting of the problem statement, the distributed data may change between two consecutive evaluations of a top-k query affecting its result. One solution to address this problem is to assume that the distributed data notifies changes to the engine that has to answer the query. First of all, it is important to note that MinTopk assumes distinct arrivals, so, it cannot be applied if the changed object has been already processed in the current window. The first contribution of this chapter is, therefore, an extension of the MinTopk algorithm (named MinTopk+) to consider indistinct arrivals of objects.

If the changed object exists in the super-top-k list, first the old object is removed from the super-top-k list, and then the object with the new score is added to the super-top-k list. If the changed object is not in the list of top-k predicted results, then it is considered as a new arrival object and check if, with a new score, it could be inserted in the top-k list. This second case is not feasible in practice, as it requires to store the value of the scoring variable x_S for all the streaming data that entered the current window, while the goal of MinTopk is to discard all streaming data that does not fit in the predicted top-k results of the active windows.

However, it is needed to generate a new score for the changed object, even if the streaming value is forgotten. Having to inspect all the streaming data entering the current window, the minimum value of the scoring variable x_S that has been seen while processing the current window is kept. Let us denote it as $min.score_S$.

Therefore, an approximated score for the changed object can be generated using $min.score_S$ as the streaming score of the changed object. As the scoring variable of the changed object cannot be greater than $min.score_S$, the generated new score is a lower bound for the real new score. In this way, the exact result in terms of $precision@k$ can be reported for the current window, but an approximated result may be reported for future evaluations.

As the scoring variable of all arrival objects in the current window are not kept, MinTopk+ is not depended on the size of the data in the window, and a subset of data are enough for top-k query answering. Further elaboration on this idea comes in Sects. 6.4.2 and 6.4.3 that, respectively, formalize how the $min.score_S$ is computed and study the memory and time complexity of a generalized version of this algorithm.

6.3.2   Updating Minimal Top-K+N Candidate List

Considering the changes in the distributed data, which affect the top-k result, this section proposes an approach that always gives the correct answer in the current window and, in some limited cases, may give an approximated answer in future windows.

The authors in [19] proposed MTK set which is necessary and sufficient for evaluating continuous top-k query. Considering changes of the objects and keeping N additional objects, the MTK set is extended and *Minimal Top-K+N Candidate list (MTK+N list)* is introduced. MTK+N list keeps K+N ordered objects that are necessary to generate top-k result. The following analysis shows that MTK+N list is also sufficient for generating correct result in the current window for most of the cases.

Assume that there exist N changes per evaluation in the distributed data, and K+N objects are kept for each window in the predicted result. Each MTK+N list consists of two areas. Let us name them K-list and N-list. Therefore, each object can be placed in 3 different areas: K-list, N-list, and outside (i.e. outside the MTK+N list).

It is worth to note that each object can be placed in different areas in different MTK+N lists.

The position of the object can change between those areas due to changes to the values assumed by the scoring variables $x_{\mathcal{D}}$ in the distributed data. Depending on the initial and the destination areas of each object, the result may be exact or approximated in each window. The following theorems analyze different scenarios for each window separately, and assume that (*i*) the previous results are correct, (*ii*) there exist N changes per evaluation in the distributed data, and (*iii*) K+N objects are kept for each window in the predicted result.

Theorem 6.1 *If the changed object is in the K-list, or the N-list and remains in one of them, or if the changed object is initially outside of the MTK+N list and remains outside, the correct top-k result can be reported for the corresponding window (current, or future).*

Proof If the changed object exists in the MTK+N list, the previous score of the object exists. The new score can only change the place of the object in the list. If the changed object is outside of the list and remains outside, no modification is needed in the MTK+N list. In both cases, the result is correct.   □

Theorem 6.2 *If the changed object was in the K-list, or the N-list, and the new score removes it from the MTK+N list, the correct top-k result can be reported for the corresponding window.*

Proof If the changed object o exists in the MTK+N list, but the new score is less than the lowest score in the MTK+N list, the object has to be removed from MTK+N list. As all the objects in the K-list are placed correctly, the exact result exists for the current window. However, after removing it, there exists one empty position in MTK+N list. If there are not any other objects in the MTK+N lists of future windows, which fit into the current MTK+N list, only object o can be added back with the new score. In previous evaluations, there may be another object with a higher score comparing to the new one of o, but it did not satisfy the constraints to be in the MTK+N list at that point in time, and it was discarded. When this case happens, the forgotten object is misplaced by object o. If during the evaluation of the window, the misplaced object o comes up in the K-list, the result is not correct.   □

Theorem 6.3 *If the changed object initially is outside the MTK+N list, and, after the changes, it moves in the MTK+N list, the result may have an approximation for the corresponding window.*

Proof When the changed object o is not in the MTK+N list, the previous information of the object in the streaming data is not accessible, appearing in the streaming data or not, and if yes, the value of scoring variable x_S is unknown. To solve this problem

Table 6.1 Summary of scenarios in handling changes

		Initial area		
		K-list	N-list	Outside
Destination area	K-list	V	V	$V^{precision@k}$, $\approx nDCG@k$
	N-list	V	V	$\approx$
	outside	$V,\approx$	$V,\approx$	V

two different approaches can be considered: first, just ignoring the changed object o which is not in the MTK+N list, second, keeping pointers to the objects that come in the streaming data in each window and keeping the minimum score of them as $min.score_S$.

Focusing on the second approach, an approximated score for o can be generated using $F(min.score_S, o.score_D)$ and the changed value of scoring variable in distributed data, which is the minimum threshold for the real score. The changed object may fit in different areas:

1. If it moves in the K-list, as the new score is a minimum threshold for real score, the real score of the object will also put it in the K-list. However, being the approximated score a lower bound, the real score may position it in a higher ranked place. So, considering $precision@k$, the result is exact, while considering $nDCG@k$, the result may be approximated.
2. If it moves in the N-list, as the new score is a minimum threshold, the real score of the object may put it in the K-list, so the result is approximated for the window.

$\square$

Table 6.1 summarizes all the explained scenarios. Assuming that the results are exact up to the current time, each cell shows the correctness of the top-k result as a function of the initial and destination areas of the changed object. The exact result is indicated by V, while the approximation in the result is shown by $\approx$. $precision@k$ and $nDCG@k$ shows the metrics used for comparing the real result with the correct one.

Theoretically, introducing another area, between N and the outside areas, can increase the correctness of the result and avoid approximation for the upcoming future windows. Considering the size of this new area equal to N, the result of the next window will also be correct for all scenarios. But, practically, the result of the experiments in Sect. 6.5 shows that keeping more objects in the MTK+N list after a certain point does not lead to a more accurate result.

Table 6.2 List of symbols used in the algorithms

Symbol	Description
MTK+N	Minimal Top-K+N list of objects
Super-MTK+N	Compact representation for MTK+N lists of objects for all active windows
O_i	An arriving object
$O_i.t$	Arriving time of object O_i
$O_i.w.start$	Starting window mark of O_i
$O_i.w.end$	Ending window mark of O_i
$O_i.score$	Score of object O_i
$w_i.lbp$	The lower bound pointer of w_i which points to the object with smallest score in the window w_i
LBP	Set of lower bound pointers for all windows that have top k objects in Super-MTK+N list
$O_{w_i.lbp}$	Object pointed by $w_i.lbp$
$w_i.tkc$	The number of items in top-k result of window w_i
W_{act}	List of active windows which contain current time in their duration
$O_{minScore}$	The object with smallest score in the Super-MTK+N list
MTK+N $.size$	Size of MTK+N list which is equal to K+N
w_{max}	Maximum number of windows
w_{exp}	The window just expired
$min.score_8$	Minimum value of scoring variable x_8 seen on the data stream while processing the current window

6.3.3 Super-MTK+N List

When a query is expressed on a sliding window, the predicted top-k results of the current and future windows partially overlap. So there exist objects which are repeated in the MTK+N lists of the current and future windows. In order to minimize the memory usage, a single integrated list for all active windows can be used instead of various MTK+N lists.

Therefore, the Super-MTK+N list is defined that consists of several MTK+N lists of all active windows (current and future). The objects in Super-MTK+N list are ordered based on their scores. In order to distinguish the top-k result of each window, for each object the starting and ending window marks are defined. The marks of each object show the period in which it is in the predicted top-k result.

Figure 6.3 shows an example of data items in a stream and the corresponding Super-MTK+N list. Figure 6.3a shows a portion of a stream between time 4 and 13, and in particular, it shows the content of window W_1 that opens at 4 and closes at

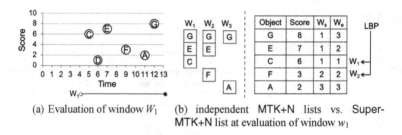

(a) Evaluation of window W_1 (b) independent MTK+N lists vs. Super-MTK+N list at evaluation of window w_1

Fig. 6.3 The example that shows the objects in top-k result after join clause evaluation of window W_1

13 (included). During window W_1, items C, D, E, F, G, and A come to the system. Consider the above example, when asked to report the top-1 objects for each window. Assuming N equal to 2, Fig. 6.3b shows the top-3 result for each window on the left. Object G is in the K-lists of all MTK+N lists, and the remaining objects are placed in the N-lists. The right side of Fig. 6.3b shows the content of Super-MTK+N list at the evaluation of window W_1. The Super-MTK+N list contains the integrated lists of all top-3 results. Objects are sorted based on their score. W_s, and W_e are starting- and ending-window-marks, respectively. The *lbp* of W_1, and W_2 are available, as those windows have top 3 objects in their predicted results.

6.3.4   Topk+N Algorithm

The previous section reports on how to extend the integrated data structure MTK list from [19] and introduces the Super-MTK+N list to handle changes in the distributed data. This section describes the Topk+N *algorithm* that evaluates top-k queries over streaming and evolving distributed data. Table 6.2 contains the description of symbols used in Algorithms 6.1, 6.2. 6.3, and 6.4.

The evaluation of a continuous top-k query over a sliding window needs to handle the arrival of new objects in the stream and the removal of old objects in the expired window. In addition to the state-of-the-art approach [19], in this problem setting the distributed data may change. So, those changes need to be handled during query processing. The proposed algorithm consists of three main steps: expiration handling, insertion handling, and change handling.

Algorithm 6.1 shows the pseudo-code of the Topk+N algorithm which gets the data stream S, the distributed data BKG, the scoring function F, and the window W as inputs and generates the top-k result for each window. In the beginning, it initializes the evaluation time. For every new arrival object O_i, in the first step, it checks if any new window has to be added to the active window list (Line 4). The algorithm keeps all the active windows in a list named W_{act}. In the next step, it checks if the time of arrival is less than the next evaluation time (i.e., the ending time of the

---

Algorithm 6.1: The pseudo-code of the proposed algorithm

Data: S, BKG, F, W

```
 1 begin
 2 │   time ← starting time of evaluation based on W ;
 3 │   foreach new object Oᵢ in the stream S do
 4 │   │   CheckNewActiveWindow (Oᵢ.t) ;
 5 │   │   if Oᵢ.t ≤ time then
 6 │   │   │   UpdateMTKN(Oᵢ) ;
 7 │   │   end
 8 │   │   else
 9 │   │   │   changedObjects ← receive changed objects from distributed dataset BKG ;
10 │   │   │   UpdateChangedObjects ( changedObjects ) ;
11 │   │   │   Get top-k result from Super-MTK+N list and generate query answer ;
12 │   │   │   PurgeExpiredWindow() ;
13 │   │   │   time ← next evaluation time ;
14 │   │   end
15 │   end
16 end
17
18 Function PurgeExpiredWindow()
19 │   i ← 0 ;
20 │   foreach O from top of Super-MTK+N list do
21 │   │   if O.w.start == wₑₓₚ then
22 │   │   │   O.w.start ++ ;
23 │   │   │   i ++ ;
24 │   │   end
25 │   │   if O.w.end < O.w.start then
26 │   │   │   Remove O from Super-MTK+N list ;
27 │   │   │   update LBP ;
28 │   │   end
29 │   │   if i == k then
30 │   │   │   break ;
31 │   │   end
32 │   end
33 │   Remove wₑₓₚ from Wₐₐₜ ;
34 │   Remove pointer of wₑₓₚ from LBP ;
```

---

current window), and it updates the Super-MTK+N list if the condition is satisfied (Lines 5–7).

Otherwise, at the end of the current window, it checks for the received changes from the distributed data (Line 9). Function UpdateChangedObjects (Line 10) gets the set *changedObjects* and updates the Super-MTK+N list based on the changes. This function is the main contribution of the Topk+N algorithm comparing to the MinTopk algorithm [19]. Getting the top-k result from Super-MTK+N list, the algorithm generates the query result (Line 11). Finally, it purges the expired window and goes to the next window processing (Lines 12–13).

6.3.4.1 Expiration Handling

When a window expires, the corresponding top-k result has to be removed from the Super-MTK+N list. As the Super-MTK+N list is an integrated view of top-k results, it is not possible to simply remove the objects, and some of the top-k objects may be also in the top-k results of the future windows. The removing of these objects can be implemented by updating their window marks and increasing the starting window marks by 1 for all the objects in the top-k result of the expired window.

Function PurgeExpiredWindow (Line 18) in Algorithm 6.1 shows the pseudo-code of expiration handling. It gets the first top-k objects from the Super-MTK+N list, whose starting window mark is equal to the expired window and increases their starting window mark by 1 (Line 22). If the starting window mark becomes larger than the end window mark, the object is removed from the Super-MTK+N list. The LBP set is updated if any pointer to the deleted object exists (Lines 25–28). Finally, the expired window is removed from the Active Windows list and the LBP set (Lines 33 and 34).

Consider the example of Fig. 6.1, where window W_0 opens at 1 (excluded) and closes at 10. Assume that we are at time 10, when window W_0 is expired, and we want to report the top-3 objects as result. Figure 6.4a shows the content of Super-MTK+N list. For window expiration, the starting window marks of the objects E, C, and B have to be increased by 1. Object B is removed from the list, as its starting window mark becomes larger than the end window mark. The *lbps* of W_0 is also removed from the LBP set. Figure 6.4b shows the Super-MTK+N list after the expiration handling of window W_0.

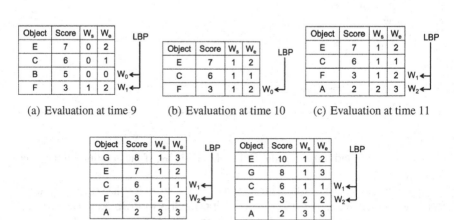

(a) Evaluation at time 9   (b) Evaluation at time 10   (c) Evaluation at time 11

(d) Evaluation at time 12   (e) Evaluation at time 13

Fig. 6.4 Super-MTK+N list content related to the example of Sect. 6.1 in different evaluation times

---

Algorithm 6.2: The pseudo-code for updating Super-MTK+N list

---

1 **Function** *UpdateMTKN(O_i)*
2   **if** $O_i.score_S < min.score_S$ **then**
3 | $min.score_S \leftarrow O_i.score_S$
4   **end**
5   **if** *Super-MTK+N list contains old version of O_i* **then**
6 | Replace O_i ;
7 | RefreshLBP() ;
8   **end**
9   **else**
10   **if** *O_i is a changed object* **then**
11 | Compute $O_i.score$ using $min.score_S$;
12   **end**
13   **else**
14 | compute $O_i.score$;
15   **end**
16   InsertToMTKN(O_i) ;
17   **end**
18 **Function** *InsertInToMTKN(O_i)*
19   **if** $O_i.score < O_{minScore}.score$ *AND all* $w_i.tkc == k$ **then**
20 | discard O_i ;
21   **end**
22   **else**
23     $O_i.w.start$ = CalculateStartWindow() ;
24     $O_i.w.end$ = CalculateEndWindow() ;
25   add O_i to MTK+N list ;
26   UpdateLBP(O_i) ;
27   **end**
28
29 **Function** *UpdateChangedObjects (Objects)*
30   **foreach** $O_i \in Objects$ **do**
31 | updateMTKN(O_i) ;
32   **end**

6.3.4.2 Handling New Arrivals and Changes

When a new object arrives in the stream, the algorithm checks if it can be added to the top-k result of the current and future windows comparing its score with the minimum score in the Super-MTK+N list. If all the conditions are satisfied, the algorithm inserts it into the Super-MTK+N list. A changed object is treated as a new arrival object. The algorithm checks if it can be added to the Super-MTK+N list. If the changed object exists in the Super-MTK+N list, it should replace the old one.

The Topk+N algorithm (see Algorithm 6.2 for the pseudo-code) updates the Super-MTK+N list based on new arriving objects on the stream S. For every object O_i in the stream, function UpdateMTKN checks if the object O_i can be inserted in the Super-MTK+N list. If the streaming score of the object is less than the value of $min.score_S$, the minimum score is updated (Lines 2–4). Keeping the minimum score allows approximating the score for changed objects as discussed in Sect. 6.3.1.

Then, it checks if the object O_i is present in the Super-MTK+N list, since Topk+N algorithm supports indistinct arrivals (different from state of the art [19]). If the Super-MTK+N list contains a stale version of O_i, the algorithm refreshes it. As the score of the replaced object O_i changes, its position in Super-MTK+N list can change too and it may move up or down in the list. Changing position in the Super-MTK+N list could affect the top-k results of some of the active windows, thus the algorithm needs to refresh the LBP set. Otherwise, when the object is not present in the Super-MTK+N list, the algorithm (1) computes the score, the starting window mark, and the ending window marks; (2) inserts the object in the Super-MTK+N list; and (3) updates the LBP set.

Algorithm 6.2 shows in more details the pseudo-code for handling the insertion of new arriving objects through the update of the Super-MTK+N list. If a stale version of the arriving object exists in Super-MTK+N list, the algorithm replaces it with the fresh one with a new score, and starting/ending window marks (Line 6). Then, the LBP set has to be refreshed based on the changes occurred in the Super-MTK+N list (Line 7). As the new value of the arriving object could change the order of objects in the Super-MTK+N list, the LBP set is recomputed. In case the object is not in the Super-MTK+N list, it computes the score, and adds the new object in the list (Line 16). If the object is a new arrival, computing the score from the values of the scoring variables is straightforward, but if object O_i is a changed object, the new score is computed getting the value of $min.score_8$ and the scoring value in the replica (Line 11), as the scoring value of all the objects are not kept (see also Sect. 6.3.1 that presents this idea).

Function InsertInToMTKN handles object insertion to the Super-MTK+N list. If the score of the object O_i is smaller than the minimum score in the Super-MTK+N list, and all active windows contain k objects as top-k result, then the arriving object is discarded (Lines 19–21). Otherwise, the future windows, in which the object can be in top-k result, are defined by computing the starting and the ending window marks (Lines 23–24). In the next step, the object is inserted into the Super-MTK+N list and the LBP set is updated (Line 26).

Function UpdateChangedObjects is used for updating the Super-MTK+N list for a set of objects, and gets the *Objects* set as input. For each object in the *Objects* set, it updates the Super-MTK+N list by refreshing the stale object in the Super-MTK+N list (Line 31).

6.3.4.3 Updating Lower Bound Pointers

As mentioned in Sect. 2.5, LBP is a set of pointers to the top-k objects with the smallest scores for all active windows that have k objects as top-k result. When a new object arrives, its score needs to be compared with those of the objects pointed by LBP for each window. If the size of any predicted top-k result for future windows is less than the size of MTK+N list (i.e. K+N), or the new object has higher score comparing to the objects that have *lbp*s, the new object can be inserted in the Super-MTK+N list.

Algorithm 6.3: The pseudo-code for updating LBP List

---

1 **Function** *UpdateLBP(O_i)*
2   **foreach** $w_i \leftarrow O_i.w.start$ **to** $O_i.w.end$ **do**
3   **if** $w_i.lbp == NULL$ **then**
4       $w_i.tkc$++;
5   **if** $w_i.tkc == MTK+N.size$ **then**
6   GenerateLBP() ;
7   **end**
8   **end**
9   **else if** $O_{w_i.lbp}.score <= O_i.score$ **then**
10       $O_{w_i.lbp}.w.start$++ ;
11   **if** $O_{w_i.lbp}.w.start > O_{w_i.lbp}.w.end$ **then**
12   Move $w_i.lbp$ by one position up in the MTK+N list ;
13   Remove $O_{w_i.lbp}$ from Super-MTK+N list;
14   **end**
15   **end**
16   **end**

After inserting the new object, the LBP set needs to be updated; in particular, those pointers that relate to the windows between the starting and the ending window marks of the inserted object. For those windows that have not got any pointer in the LBP set, the size of the top-k result is increased by 1. If the size becomes equal to k, the pointer is created for the window and added to the LBP set.

If the window has got a pointer in LBP set and the score of the inserted object is less than the score of the pointed object, then the last top-k object in the predicted result is removed from the list, so the starting window mark have to be incremented by 1. If the starting window mark becomes greater than the ending window mark for any object, the pointer moves up by one position in the Super-MTK+N list and the object is removed from the Super-MTK+N list.

Algorithm 6.3 shows the pseudo-code for updating the LBP set after inserting the new object to the Super-MTK+N list. For all the affected windows from the starting to the ending window marks of the inserted object, if the window does not have any *lbp*, the cardinality of the top-k result is incremented by 1 (Line 4). If the cardinality of the top-k result of a window reaches the K+N, Function GenerateLBP generates the pointer to the last top-k object of that window and adds it to the LBP set (Line 6).

If the window has a pointer in LBP set, the score of the inserted object is compared with the score of the pointed object (i.e. the last object in top-k result with the lowest score). If the inserted object has higher score, the last object in top-k result is removed by increasing the starting window mark by 1 (Line 10). If the starting window mark of the object becomes greater than ending window mark, the *lbp* is moved one position up in the Super-MTK+N list and the algorithm removes the object from the Super-MTK+N list (Lines 11–14).

Consider our example in Fig. 6.1 at time 11. Figure 6.4b shows the content of Super-MTK+N list after the expiration of window W_0. At time 11, object A comes to the system, and based on its score it is inserted to the Super-MTK+N list with starting window mark equal to 2 and ending window mark equal to 3. As now we have 3 objects in window W_2, the *lbp* of the window is added to the LBP set. Figure 6.4b shows the changes applied on the Super-MTK+N list.

At time 12, object G comes to the system, and it is inserted at the top of the Super-MTK+N list. As Windows W_1, and W_2 have 3 objects in their top-3 result, the algorithm has to remove the last object of each window. So the starting window marks of objects C and F need to be increased by one, and *lbp* of Windows W_1, and W_2 should be modified. Object F was the last object in the result of window W_1, and after inserting object G, object C becomes the last object and *lbp* of Windows W_1 points to it. For window W_2, the last object changes from object A to object F, and the *lbp* moves one position up accordingly.

Figure 6.4 shows how handling changes can affect the content of the Super-MTK+N list and of the top-k query result. At the evaluation time of W_1 at time 13, after handling new arrivals of window W_1, the content of the Super-MTK+N list is as in Fig. 6.4e. As the score of object E changes from 7 to 10 (Fig. 6.1c), it is considered as an arriving object with a new score, so, it is placed in the Super-MTK+N list above object G. The LBP set does not change in this case.

Comparing to the MinTopk algorithm [19], the Topk+N algorithm has the following additional futures: (*i*) it computes the minimum score on the streaming side to approximate score of changed objects, (*ii*) it handles distinctive arrival of objects, and (*iii*) it handles changed objects.

6.4 AcquaTop Solution

Using Super-MTK+N list and Topk+N algorithm, it is possible to process continuous top-k query over streaming and distributed data while getting notification of changes from the distributed data. As anticipated in Sect. 6.1, this solution works in a data center, where the entire infrastructure is under control, but it does not when there are high latency, low bandwidth, and even rate-limited access as in the two examples of Sect. 6.1. In those cases the engine, which continuously evaluates the query, has to pull the changes from the distributed data and, thus, it is at risk of violating the reactiveness requirement.

The remainder of the section presents the AcquaTop solution to address this problem. Section 6.4.1 introduces the AcquaTop Framework. Section 6.4.2 presents the details of the AcquaTop algorithm, and the proposed maintenance policies. Finally, Sect. 6.4.3 reviews the cost analysis.

6.4.1 AcquaTop Framework

As mentioned in Chap. 3, ACQUA addresses this problem by keeping a local replica of the distributed data and using several maintenance policies to refresh such a

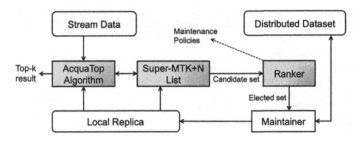

Fig. 6.5 The AcquaTop framework

replica. Considering the architectural approach presented in Fig. 3.1 as a guideline, this section presents a second solution to the problem, named AcquaTop *framework*. It keeps a local replica of the distributed data and focuses the refresh budget on the data items that have the largest potential impact on the current and future top-k answer after every evaluation.

Figure 6.5 shows the architecture of the AcquaTop framework. AcquaTop gets data from the stream and the local replica and, using the Super-MTK+N list structure, it evaluates continuous top-k query at the end of each window. The Super-MTK+N list provides the candidate set for updating. Notably, such a set is a small subset of the objects that logically should be stored in the window since the proposed approach discards objects that do not enter the predicted top-k results when they arrive. The Ranker gets the candidate set and orders its elements based on the criteria of the different maintenance policies. The maintainer gets the top γ elements, namely the Elected set, where γ is the refresh budget for updating the local replica.

When γ is enough to update all the stale elements in the replica, the AcquaTop algorithm has the same guarantees of the Topk+N algorithm (see Table 6.1 and the theorems in Sect. 6.3.2). AcquaTop *algorithm* and the proposed maintenance policies approximate the result.

6.4.2 AcquaTop Algorithm

In top-k query evaluation, after processing the new arrivals of each window, the set of objects which have been updated in the local replica is prepared by fetching a fresher version from the distributed data. Algorithm 6.4 shows the pseudo-code of the AcquaTop Algorithm for handling changes in local replica in addition to handling insertion of new arrival objects.

In the first step, the evaluation time is initialized. Then, for every new arriving objects, it checks if any new window has to be added to the active window list (Line 4). If the time of arrival is less than the next evaluation time (i.e., the ending time of the current window), it updates the Super-MTK+N list (Lines 5–7).

Algorithm 6.4: The pseudo-code of AcquaTop algorithm

---

1 **begin**
2   *time* ← starting time of evaluation ;
3   **foreach** *new object O_i in the stream S* **do**
4   CheckNewActiveWindow ($O_i.t$) ;
5   **if** $O_i.t \leq time$ **then**
6 | UpdateMTKN(O_i) ;
7   **end**
8   **else**
9   *changedObjects* ← UpdateReplica(Super-MTK+N list);
10   TopkN (*changedObjects*) ;
11   Get top-k result from Super-MTK+N list and generate query answer ;
12   PurgeExpiredWindow() ;
13   *time* ← next evaluation time ;
14   **end**
15   **end**
16 **end**
17
18 **Function** *UpdateReplica(Super-MTK+N list , policy)*
19   *electedSet* ← UpdatePolicy (*Super-MTK+N list* , policy) ;
20   **foreach** $O_i \in electedSet$ **do**
21   **if** *new value of scoring variable of O_i $\neq$ replica value of scoring variable of O_i* **then**
22   update replica for O_i ;
23   add O_i to list *changedObjects* ;
24   **end**
25   **end**
26   **return** *changedObjects* ;

At the end of the current window, Function UpdateReplica gets the Super-MTK+N list and returns the set of changed objects in the replica (Line 9). Then, Function TopkN (Line 10) gets the set *changedObjects* and updates the Super-MTK+N list based on changes. The algorithm considers changed objects as new arriving objects with different scores. It removes the stale version of the object from the Super-MTK+N list and reinserts it if the constraints are satisfied. Then, getting the top-k result from Super-MTK+N list, the algorithm returns the query answer (Line 11). Finally, it purges the expired window and goes to the next window (Lines 12–13).

Function UpdateReplica in Algorithm 6.4 updates the replica based on the content of the Super-MTK+N list and the maintenance policy. In the beginning, Function UpdatePolicy (Line 19) gets the Super-MTK+N list and the *policy* as input, and based on the input policy, it returns the *electedSet* of objects for updating. The following four sections detail the maintenance policies. For every object in the *electedSet*, if the new value of the scoring variable $x_{\mathcal{D}}$ differs from the one in replica, it updates the replica and puts the object in the set *changedObjects* (Lines 20–25). Finally, Function UpdateReplica returns the set *changedObjects*.

6.4.2.1   Top Selection Maintenance Policy (AT-TSM)

Specific maintenance policies are needed for top-k query evaluation. The intuition is straightforward: since AcquaTop algorithm makes it possible to predict the top-k result of the future windows, updating the replica for those predicted objects catches the opportunity to generate more accurate result. As a consequence, the rest of the data in the replica has less priority for updating.

The predicted top-k results of future windows are kept in the Super-MTK+N list. Based on the AcquaTop algorithm, the top-k objects of the current window have high probability to be in the top-k result of future windows. Therefore, updating the top-k objects can affect the relevance of the result of future windows more than updating objects far from the first top-k. Based on this intuition, the AT-TSM *policy* selects objects from the top of the Super-MTK+N list for updating the local replica. The proposed policy gives priority to the object with a higher rank, as it focuses on more relevant result. The hypothesis is that comparing to the other policies, AT-TSM can have a higher value of $nDCG@k$.

For example, let us assume that k is equal to 3, N is equal to 1, and γ is equal to 2. Consider the example introduced in Sect. 6.1, and the Super-MTK+N list showed in Fig. 6.4e. At the end of window W_1, AT-TSM policy selects object G and E for updating from the top of the Super-MTK+N list.

6.4.2.2   Border Selection Maintenance Policy (AT-BSM)

Super-MTK+N list contains K+N objects for each window, and each object in the predicted result is placed in one of the following areas: the K-list, which contains the top-k objects with the highest rank; or the N-list, which contains the next N items after top-k ones. AT-BSM *policy* focuses on the objects around the border of those two lists and selects objects for updating around the border.

The intuition behind AT-BSM is that objects around the border have higher chances to move between the K- and the N-list [21]. Indeed, updating those objects may affect the top-k result of future windows. The policy concentrates on the objects that may be inserted in or removed from the top-k result and can generate more accurate results. So, the hypothesis is that comparing with other policies, AT-BSM policy has higher value of $precision@k$.

For example, consider Fig. 6.4e, and assume that k is equal to 3, N is equal to 1, and γ is equal to 2. AT-BSM policy selects objects C and F for updating from the border of the K- and the N-list in the Super-MTK+N list.

6.4.2.3   All Selection Maintenance Policy (AT-ASM)

The upper bound accuracy and relevancy of AcquaTop is the case when there is no limit for the refresh budget, i.e. all the elements in the Super-MTK+N list can be updated. This policy is named AT-ASM. The hypothesis is that AT-ASM policy has the highest[2] accuracy and relevancy as it has no constraint on the number of accesses to the distributed data, and updates all the objects in the predicted top-k results. AT-ASM policy is not useful in practice, but it is used to verify the correctness of the experiments reported in Sect. 6.5. In our example of Fig. 6.4e, AT-ASM policy updates all the objects in the Super-MTK+N list.

6.4.2.4   AT-LRU and AT-WBM Policies

As an alternative, AcquaTop algorithm and Super-MTK+N list can be used to eval-uate top-k query, while applying state-of-the-art maintenance policies from ACQUA for updating the local replica. ACQUA shows that WSJ-WBM and WSJ-LRU poli-cies perform better than others while processing join queries. Combining those poli-cies with AcquaTop algorithm, the following policies are proposed: AT-LRU, and AT-WBM. The hypothesis is that AT-LRU works when most recently used objects appear in the top-k result of future windows. AT-WBM policy works when being in the top-k result is correlated with staying longer in the sliding window.

6.4.3   Cost Analysis

The memory size required for each object o_i in the Super-MTK+N list is equal to $(Object.size + 2 * Reference.size)$, as the object and its two window marks are kept in the Super-MTK+N list. Based on the analysis in [19], in the average case, the size of the super-top-k list is equal to $2K$ (K is the size of MTK set). Therefore, in the average case, the size of the Super-MTK+N list is equal to $2 * MTK+N.size = 2 * (K + N)$. Notably, *the memory complexity of the Super-MTK+N list is constant*, as the value of K and N are fixed. it depends neither on the volume of data that comes from the stream, nor on the size of the distributed data.

The CPU complexity of the proposed algorithm is computed as follows. The complexity of handling object expiration is equal to $O(MTK+N.size)$, as it is needed to go through the MTK+N list to find the first k objects of the just expired window.

For handling the new arrival object, the cost for each object is:

$$P * (log(MTK + N.size) + W_{act}.size + C_{aaw} + C_{albp}) + \qquad (6.2)$$
$$(1 - P) * (1 + W_{act}.size),$$

where P is the probability that object o_i will be inserted in the Super-MTK+N list, C_{aaw} is the number of affected active window, C_{albp} is the number of affected pointers in LBP set, and $W_{act}.size$ is the size of the active window list.

---

[2]As noted in Sect. 6.3.2, and specifically in Table 6.1, updating all elements in the MTK+N list is not sufficient to guarantee correctness.

If the probability of inserting object o_i in the Super-MTK+N list is P, the cost for positioning it in the Super-MTK+N list is equal to $log(MTK+N.size)$ by using a tree-based structure for storing the Super-MTK+N list. The cost of computing the starting window marks is equal to $W_{act}.size$, as all the active windows must be checked as a candidate. The cost of updating the counters of all affected active windows is C_{aaw}, and the cost of updating all affected pointers in LBP set is C_{albp}.

With probability $(1 - P)$, the object is discarded with the cost of one single check with the lowest score in Super-MTK+N list and $W_{act}.size$ checks of active window counters.

For handling the changed object, the cost for each object is:

$$2 * log(MTK+N.size) + O(MTK+N.size), \qquad (6.3)$$

where $2 * log(MTK+N.size)$ is the cost of removing the old object and inserting it with a new score, and $O(MTK+N.size)$ is the cost of refreshing the LBP set.

Therefore, in the average case, the CPU complexity of the proposed algorithm is $O(N_{new} * (log(K + N) + W_{act}.size) + N_{changes} * (K + N))$. The analysis shows that the most important factors, in CPU cost of AcquaTop algorithm, are the size of MTK+N and the number of active windows (i.e. $W_{act}.size$), which are fixed during the query evaluation. Therefore, the *CPU cost is constant* as it is independent from the size of the distributed data and the rate of arrival objects in the data stream.

Based on these cost computations, the proposed approach can be compared with the state-of-the-art ones. Comparing to the MinTopk algorithm, AcquaTop has memory overhead equal to $O(N)$, but N can be set to 0 if the distributed data does not change. As stated before, the computational cost of AcquaTop algorithm is equal to $2 * log(MTK+N.size) + O(MTK+N.size)$, while for the MinTopk algorithm, the cost is equal to $2 * log(MTK.size)$. So even when $N=0$, AcquaTop still has a small constant overhead in the worst case.

Comparing to ACQUA, AcquaTop's memory cost is equal to $O(MTK+N.size)$, while ACQUA has to keep all data items that come in the window to compute the top-k result of the window. Moreover, the state-of-the-art shows that using a materialization-then-sort approach (like ACQUA) has higher computational overhead comparing to the incremental approaches (MinTopk and AcquaTop).

6.5   Evaluation

This section reports the results of the experiments that evaluate the proposed policies. Section 6.5.1 introduces the experimental setting. Section 6.5.2 shows the result of the preliminary experiment. In Sect. 6.5.3, the research hypotheses are formulated. The rest of the sections report on the evaluation of the research hypotheses.

Fig. 6.6 Data stream characteristics

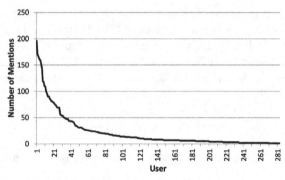

(a) distribution of number of mentions

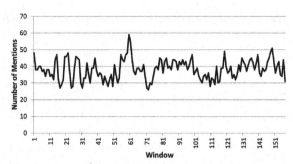

(b) Number of mentions per window

6.5.1 Experimental Setting

As an experimental environment, an Intel i7@1.7 GHz with 8 GB memory and a SSD disk are used. The operating system is Mac OS X 10.13.2 and Java 1.8.0_91 is installed on the machine. The experiments performed by extending the experimental setting of [2].

The experimental data consists of streaming and distributed data. The streaming data contains tweets mentioning 400 verified users of Twitter. The data is collected by using the streaming API of Twitter for around three hours of tweets (9462 s). As one can expect, the number of mentions per user during the evaluation has a long-tail distribution, in which few users have high number of mentions, and most of the users have little mentions. The profiling of the number of mentions per window shows min = 26, median = 38, and max = 59. Figure 6.6a shows the distribution of the number of mentions, and Fig. 6.6b shows the number of mentions per window.

The distributed data was fetched from twitter's REST APIs recordings every minute for each user the number of followers. Differently from the example in Listing 6.1, to better resemble the problem presented in Sect. 6.1, for each user u and each minute i, the difference between the number of followers at i (nf_i) and the one at the previous minute i-1 (nf_{i-1}) is added to the distributed data. Let us denote such a

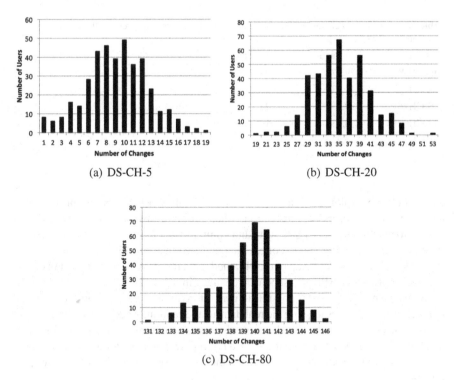

Fig. 6.7 Distribution of number of users per number of changes

difference with dfc_i. It holds that $dfc_i = nf_i - nf_{i-1}$. So, the top-k query presented in Sect. 6.1 is modified for the experimental evaluation.

The length of the window is set equal to 100 s, and the slide equal to 60 s. 150 iterations of the query evaluation run (i.e. there are 150 slided windows for the recorded period of data from twitter) to compare different maintenance policies. The scoring function for each user is a linear combination of the number of mentions (named mn) in the streaming data and the value of dfc in the distributed data. Notably the values of mn, and dfc increase or decrease during the iterations, but the selected linear scoring function is monotonic as assumed in top-k query literature. The scoring function computes the score as follows:

$$score = F(mn, dfc) = w_s * norm(mn) + w_d * norm(dfc), \qquad (6.4)$$

where, $norm$ is a function that computes the normalized value of its input, considering the minimum and maximum value in the input range, w_s is the weight used for the number of mentions, and w_d in the weight used for the number of followers.

For experimental evaluation, it is needed to control the average number of changes in the distributed data. Before controlling it, the distribution of changes in the recorded

Table 6.3 Summary of characteristics of the distributed datasets which reports the statistic related to the number of changes per invocation

Dataset	Average	Median	1st quartile	3rd quartile
DS-CH-80	79.97	94	77	96
DS-CH-40	40.33	47	40	48
DS-CH-20	20.45	23	20.5	24
DS-CH-10	10.33	12	10	12
DS-CH-5	5.53	6	5	6

data should be explored. Notably, Twitter APIs allow asking for the profile of a maximum of 100 users per invocation,[3] thus multiple invocations are needed per minute to get the number of followers and compute the dfc for each of the 400 users. In total, 702 invocations run to collect the data used over the 150 iterations.

Exploring the characteristic of the obtained data, it is possible to note that in every invocation of twitter API, 80 users have changes in dfc.

With this information, it is possible to generate synthetic datasets with a decreased number of changes by sampling the real dataset and randomly decreasing the average number of changes in dfc. To decrease the average number of changes, for each invocation, users who have changes in dfc, are selected randomly, and set their value of dfc to the previous value to reach the target average number of changes per invocation.

It is also easy to determine that this method introduces many ties in the scores, which as known in top-k query answering literature simplifies the problem. In order to keep the problem as hard as possible, the changes in dfc are altered by adding random noise.

Applying those methods, five datasets are generated in which there are on average 5, 10, 20, 40, and 80 changes in each invocation. The value 80 is the maximum number of changes in each invocation for the dataset. So, the dataset with 80 changes in each invocation is the extreme test case in the evaluations. Each generated dataset has a normal distribution of number of changes. Each dataset has different mean value, but there is not any significant difference between variances. Figure 6.7 shows the distribution for three of those datasets. They are synthetic, but realistic.

In order to reduce the risk of introducing a bias in synthetic data generation, for each number of changes, a test case is produced that contains five different datasets. In the remainder of the chapter, the notation DS-CH-x is used to refer collectively to the five datasets whose average number of changes per invocation is equal to x. Table 6.3 shows the characteristics of the generated datasets.

---

[3] Twitter API returns the information of up to 100 users per request, https://developer.twitter.com/en/docs/accounts-and-users/follow-search-get-users/api-reference/get-users-lookup.

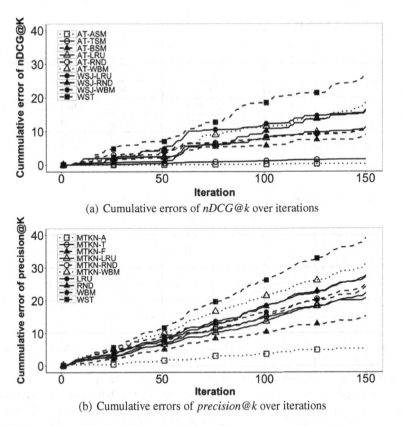

(a) Cumulative errors of *nDCG@k* over iterations

(b) Cumulative errors of *precision@k* over iterations

Fig. 6.8 Result of Preliminary Experiment

6.5.2 Preliminary Experiment

This experiment checks the relevancy and accuracy of the top-k result for all the maintenance policies over 150 iterations. DS-CH-20 test case is selected for this experiment. In the first step, the total result is checked for each iteration: on average there are 30 items in the query result. Therefore, the default value for K set equal to 5, which is around 15% of the average size of the total result. The refresh budget is set equal to 7, so theoretically, there is enough budget to refresh all the answers of top-k query.

In order to set a default value for parameter N, the distributed data has to be analyzed. As stated in Sect. 6.5.1, during 9462 s of recording data from twitter API, there are 702 invocations. Therefore, on average there are 7.42 invocations per window with 100 s length ($702 \div 9462 \times 100 = 7.42$). Based on the definition, in DS-CH-20 test case, there are 20 changes per invocation on average. So, the average number of changes per window is equal to $7.42 \times 20 = 148.4$. Considering that there exist 400 users in total and 30 users on average in the result set, there are 11.13

Table 6.4 Parameter Grid

Parameter	(Default) values	Description
CH	(20) {5,10,20,40,80}	Average Number of changes per invocation
B	(7) {1,3,5,7,10,15,20,25,30}	Refresh budget
K	(5) {5,7,10,15,20,30}	Number of top-k result
N	(10) {0,10,20,30,40}	Number of additional elements in MTK+N list

Table 6.5 Summary of experiments

Experiment	Hypothesis	B	CH	K	N
0	–	7	20	5	20
1	HP.4.1	B	10	5	10
2	HP.4.2	7	CH	5	10
3	HP.4.3	7, 15	10	K	10
4	HP.4.4	7, 15	10	5	N

changes per window ($\frac{148.4}{400} \times 30 = 11.13$). So, the default value of N is set equal to 10 for the MTK+N list.

In order to investigate the hypotheses, an Oracle at each iteration i provides the correct answers $Ans(Oracle_i)$, which is compared with the possibly erroneous ones of the query, $Ans(Q_i)$. Given that the answers are ordered lists of the users' IDs, the following metrics are used: $nDCG@k$ and $precision@k$. Moreover, the *cumulated $nDCG@k$* and $precision@k$ are considered. The cumulated $nDCG@k$ ($precision@k$) is defined at the Jth iteration as follows:

$$nDCG@k^C(J) = \sum_{i=1}^{J} nDCG@k(Ans(Q_i), Ans(Oracle_i)) \qquad (6.5)$$

$$precision@K^C(J) = \sum_{i=1}^{J} precision@K(Ans(Q_i), Ans(Oracle_i)) \qquad (6.6)$$

where the $nDCG@k$ of the iteration i is denoted as $nDCG@k(Ans(Q_i), Ans(Oracle_i))$ and the $precision@k$ of the iteration i as $precision@k(Ans(Q_i), Ans(Oracle_i))$. Higher value of $nDCG@k$ and $precision@k$ show more relevancy and accuracy of the result set, respectively.

The query evaluation run for 150 iterations for each policy and the cumulative error related to $nDCG@k$ and $precision@k$ metrics are computed for every iteration. Figure 6.8 shows the result of the experiment. At the beginning (iteration 1 to 50) it is difficult to identify policies with better performance, but while the iteration number increases, distinct lines become detectable and comparison between different policies becomes easier. Therefore, for the rest of the experiment the value of $nDCG@k^C(150)$, or $precision@k^C(150)$ are considered for comparing the relevancy and accuracy of different policies. Abusing notation, in the rest of the chapter, $nDCG@k$, or $precision@k$ is used for referring to them.

6.5.3 Research Hypotheses

Table 6.4 describes various dimensions of the space, in which the hypotheses are formulated, and shows the values for each parameter that is used in the experiments.

Two maintenance policies (WST, and WSJ-RND) are used as a baseline to compare proposed policies with. WST maintenance policy is a lower bound w.r.t. the comparison in terms of accuracy with the Oracle. WSJ-RND randomly selects objects for updating from the Candidate set. It is expected that the proposed policies outperform WSJ-RND policy. As said in Sect. 6.4.2.3, AT-ASM policy is also introduced as an upper bound. It is expected that all policies show worse accuracy and relevancy than AT-ASM w.r.t. the Oracle.

In addition to the baseline policies, WSJ-LRU, and WSJ-WBM policies are considered from ACQUA in order to compare them with the proposed policies. Moreover, as mentioned in Sect. 6.4.2.4, AT-LRU, and AT-WBM are introduced, which use AcquaTop algorithm and Super-MTK+N list, while applying maintenance policies from ACQUA for updating the local replica.

In general, the hypothesis said that the proposed policies outperform the ACQUA policies within the setting of this chapter. As AcquaTop algorithm only keeps the objects which can participate in the top-k result and discards the rest of the data stream, even comparable results with the ACQUA policies (WSJ-WBM, and WSJ-LRU) are good. Indeed, AcquaTop algorithm has significant optimization in memory usage, while ACQUA's memory complexity depends on the amount of data in the window, AcquaTop framework's memory only depends on K, and N.

The hypotheses are formulated as follows:

Hp.4.1 For every refresh budget the proposed policies (AT-TSM, and AT-BSM) report more relevant (accurate) or comparable results with the ACQUA policies.

Hp.4.2 For datasets with different average number of changes per invocation (CH) the proposed policies generate more relevant (accurate) or comparable results with the ACQUA policies.

Hp.4.3 Considering enough refresh budget for updating the replica, for every value of k the proposed policies report more relevant (accurate) or comparable results with the ACQUA policies.

Hp.4.4 Considering enough refresh budget for updating replica, for every value of $N \geq CH$ the proposed policies report more relevant (accurate) or comparable results with the ACQUA policies.

Table 6.5 summarizes a significant subset of the experiments. In each experiment, one parameter has various values and the rest of them have a default value. For every experiment $nDCG@k$, and $precision@k$ are computed to compare the relevancy and accuracy of the generated results w.r.t. the Oracle.

6.5.4   Experiment 1—Sensitivity to the Refresh Budget

This experiment checks the sensitivity to the refresh budget for different policies to test Hypothesis Hp.4.1 As mentioned in Sect. 6.5.2, based on the analysis of the data stream and the distributed data, K is set equal to 5, and N equal to 10. The experiment run over the DS-CH-20 test case for different refresh budgets ($\gamma \in \{1, 3, 7, 10, 15, 20, 25\}$), considering $\gamma = 1$ as the extreme case where there is minimum budget for updating, and $\gamma = 25$ for maximum budget that should allow to refresh all stale elements, as there are on average $2 * (K + N) = 30$ items in the MTK+N list (see Sect. 6.4.3).

Figure 6.9 shows the result of the experiment for different budgets. Figure 6.9a shows the median of cumulative $nDCG@k$ with error bars over five datasets for different policies and refresh budgets. Y axis shows the value of cumulative $nDCG@k$. The maximum value on $nDCG@k$ is equal to 150, because in each iteration the maximum value of $nDCG@k$ is equal to 1 for the correct answer and there are 150 iterations. X axis shows different values of refresh budget and each line identifies a maintenance policy. Figure 6.9b shows the median of cumulative $precision@k$ with error bars in the same way.

Figure 6.9a shows that **AT-ASM** has the highest relevancy in top-k results as it updates all the objects in the **Super-MTK+N** list without considering the refresh budget. **WST** policy is also not sensitive to refresh budget as it does not update the local replica. Therefore, low relevancy of the result is expected for **WST** policy. When there is a small refresh budget for updating local replica, the proposed policies (**AT-TSM**, and **AT-BSM**) perform like other policies and have the same (extremely low) relevancy in top-k result. But, when there is enough refresh budget (i.e., 3 to 15), **AT-TSM**, and **AT-BSM** policy outperform other policies. When the value of the refresh budget is high ($\gamma \geq 20$), **AT-LRU** is as good as **AT-ASM**, **AT-TSM**, and **AT-BSM** policies in relevancy. This is expected because considering K=5 and N=10, MTK+N size is equal to 15 and based on [19], there are $2 \times 15 = 30$ objects in Super-MTK+N list on average. So, for refresh budget near to 30, almost the entire Super-MTK+N list is refreshed.

Figure 6.9b shows the accuracy of the top-k results. Like the chart of Fig. 6.9a, **AT-ASM** and **WST** policies are not sensitive to refresh budget. **AT-BSM** policy outperforms other policies for most of the refresh budgets ($\gamma \leq 20$). For low refresh budgets ($\gamma < 7$) **AT-TSM** can generate top-k results as accurate as others, but for

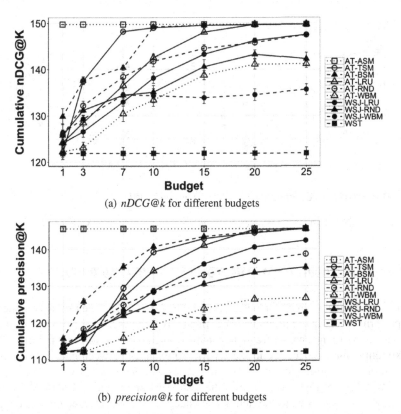

(a) *nDCG@k* for different budgets

(b) *precision@k* for different budgets

Fig. 6.9 Result of Experiment 1—Relevancy and Accuracy for different value of refresh budget

budgets between 7 to 15 it has higher accuracy comparing to other policies except AT-BSM policy. For large budgets, AT-TSM, AT-BSM, and AT-LRU are as good as AT-ASM.

Figure 6.9 shows some elbow points around refresh budget equal to 10 or 15, in which the performance starts to rise slowly. The experiment was done over the DS-CH-20 test case, and the computation in Sect. 6.5.2 shows that on average there exist 11.13 changes in the result set. Therefore having a high number of refresh budget ($\gamma \geq 15$) does not help to improve the performance, as there are lower numbers of changes comparing to the refresh budget.

From a practical perspective, this analysis confirms what said in Sect. 6.3: if there exists enough refresh budget for updating the top-k result, AT-TSM policy is the best option. Applying AT-TSM policy more relevant data can be achieved, while the rank position of each data item is important. AT-BSM policy outperforms others when only accuracy is considered.

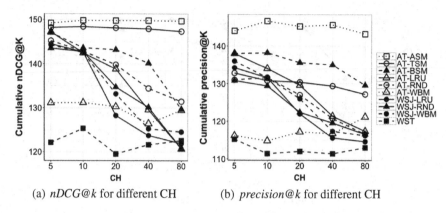

(a) *nDCG@k* for different CH   (b) *precision@k* for different CH

Fig. 6.10 Result of Experiment 2—Relevancy and Accuracy for different value of CH

6.5.5   Experiment 2—Sensitivity to Change Frequency (CH)

In this experiment, refresh budget is set equal to 7, i.e., where the proposed policies outperform others in the previous experiment. Hypothesis Hp.4.2 is tested to check the sensitivity to the change frequency in distributed data for different policies. The maximum change frequency in the data is 80 changes between each consecutive evaluation. The top-k query runs over test cases with various CH values, setting N to 10, and K to 5. Figure 6.10 shows the result of Experiment 2. Charts show that AT-TSM is not sensitive to the number of changes, as both relevancy and accuracy of the result do not show any noticeable variation.

Figure 6.10a shows the relevancy of the result considering *nDCG@k* metric for different CH. For most of the policies, while there is fewer number of changes in the data, the relevancy is higher. Both AT-TSM and AT-BSM policies outperform others.

Figure 6.10b shows the accuracy of the top-k result for various CH considering *precision@k* metric. In most of the policies, increasing the number of changes reduces the accuracy of the result. AT-BSM generates the more accurate top-k result, considering *precision@k*. AT-TSM has almost the same accuracy for all CH, but in AT-BSM the accuracy decreases for high CH. The robust performance of the AT-TSM policy for different CH is not expected. Theoretically, for the higher value of CH, it is needed to keep more objects in the Super-MTK+N list (i.e., $N \simeq CH$), but practically AT-TSM policy has almost the same relevancy and accuracy for different values of CH.

6.5.6   Experiment 3—Sensitivity to K

The result of Experiment 1 shows that, for refresh budget between 7 and 15, AT-TSM, and AT-BSM policies outperform other policies both in relevancy and in accuracy.

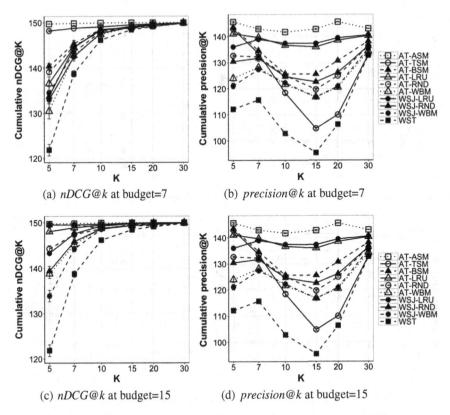

(a) *nDCG@k* at budget=7 (b) *precision@k* at budget=7

(c) *nDCG@k* at budget=15 (d) *precision@k* at budget=15

Fig. 6.11 Result of Experiment 3—Relevancy and Accuracy for different value of K

So, in this experiment, some extreme conditions are explored. Focusing on the middle area of budget selection, the refresh budget is set equal to 7 and 15, which are the minimum and maximum refresh budgets in this area respectively. The query runs for different values of K (i.e. $K \in \{5, 7, 10, 15, 20, 30\}$) to test Hypothesis Hp.4.3. Using the Oracle it is possible to note that the average of result items in each window is equal to 31.9, so $K = 30$ is considered as the maximum value. Notably, when K is 100% of the window content the policies are tested in the ACQUA setting.

Figure 6.11a, c shows that for different K, AT-TSM, and AT-BSM perform better than others and the results are more relevant. They also generate more relevant result while the refresh budget is higher ($\gamma = 15$).

Figure 6.11b, d shows that for low values of K, (i.e. $K < 7$), AT-TSM, and AT-BSM perform better than others. When refresh budget is equal to 7, and $K \geq 7$, most of the policies outperform AT-TSM and AT-BSM, and WSJ-LRU, and AT-LRU are the best policies. When the refresh budget is equal to 15 and $K \geq 7$, in general, the results are more accurate, and WSJ-LRU and AT-LRU are the best policies after AT-ASM. AT-BSM is better than the remaining policies, while AT-TSM is the worst policy (even after WST).

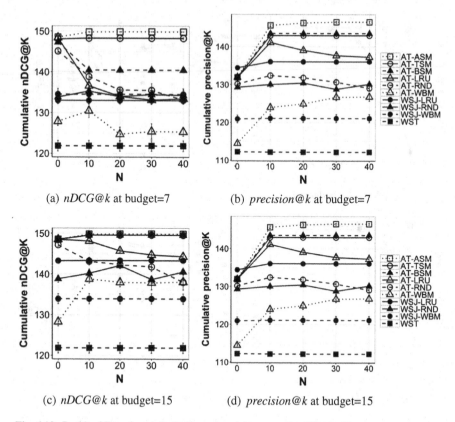

(a) *nDCG@k* at budget=7   (b) *precision@k* at budget=7

(c) *nDCG@k* at budget=15   (d) *precision@k* at budget=15

Fig. 6.12 Result of Experiment 4—Relevancy and Accuracy for different N

Unexpectedly, the lesson learned from observation is that focusing on a specific part of the result (e.g. top of the result) and trying to update that part could generate more errors when the refresh budget is not enough to update the entire top-k result (i.e., $\gamma < K$). In this case, selecting from all the objects in the window or Super-MTK+N list, as done in WSJ-LRU, or AT-LRU, can lead to more accurate results.

6.5.7 Experiment 4—Sensitivity to N

In this experiment, focusing on the middle area of Fig. 6.9, in which AT-TSM, and AT-BSM policies outperform other policies both in relevancy and in accuracy, the refresh budget set equal to 7 and 15, which are the minimum and maximum refresh budgets in this area. The query runs for different N (i.e. N∈ {0, 10, 20, 30, 40}) to test Hypothesis Hp.4.4. Notice that Hp.4.4 should be confirmed for large N while small N are extreme situations where Hp.4.4 may not hold.

Figure 6.12 shows that **AT-TSM**, and **AT-BSM** policies perform better than others. **AT-TSM** policy has higher relevant results considering $nDCG@k$, while **AT-BSM** generates more accurate results considering $precision@k$. This observation gives us an insight. Focusing on the top result can lead to a more relevant result, while focusing on the border of the K and the N area, can give us a more accurate result.

Comparing the plots in Fig. 6.12, it is found that giving more refresh budget, it is possible to fill the gap between **AT-TSM**, and **AT-BSM** with **AT-ASM** and generate more relevant and accurate results.

Theoretically, keeping additional N objects in **Super-MTK+N** list leads to more relevant and accurate results. Figure 6.12 also shows that **AT-ASM** policy performs better when there are higher values of N. However, from a practical perspective, if there is not enough refresh budget to update the replica, more relevant and accurate results cannot be generated.

6.5.8 Wrap up

Tables 6.6, and 6.7 summarize the improvements of proposed policies (**AT-TSM**, and **AT-BSM**) comparing them with the ACQUA ones (**WSJ-WBM**, and **WSJ-LRU**) considering $nDCG@k$, and $precision@k$. For each experiment (1 to 4), policies are compared and the maximum and minimum improvement for both metrics are found.

Table 6.6 Improvement of proposed policies (in percentage) comparing with the state-of-the-art ones in different experiments considering $nDCG@k$

Experiment	1-Refresh budget		2-CH		3-K		4-N	
Comparison	max	min	max	min	max	min	max	min
AT-TSM vs WSJ-WBM	11.68	−3.58	18.28	2.75	11.68	0.2	11.71	10.05
AT-TSM vs WSJ-LRU	14.22	−1.76	20.87	2.57	11.46	0.07	11.46	3.65
AT-BSM vs WSJ-WBM	11.7	2.76	11.86	0.75	11.7	−0.29	11.72	4.39
AT-BSM vs WSJ-LRU	9.1	0.68	13.27	0.65	5.58	−0.34	10.71	3.65

Table 6.7 Improvement of proposed policies (in percentage) comparing with the state-of-the-art ones in different experiments considering $precision@k$

Experiment	1-Refresh budget		2-CH		3-K		4-N	
Comparison	max	min	max	min	max	min	max	min
AT-TSM vs WSJ-WBM	19.16	−3.76	10.22	−2.35	18.04	−10.26	18.3	9.06
AT-TSM vs WSJ-LRU	8.27	−2.69	11.94	−1.04	5.06	−23.72	5.29	−1.87
AT-BSM vs WSJ-WBM	19.39	1.44	14.99	1.47	18.5	2.8	18.5	5.47
AT-BSM vs WSJ-LRU	11.95	0.55	16.78	2.83	5.47	−8.59	9.06	−1.87

Table 6.8 Summary of the verification of the hypotheses. Overall, AT-TSM shows better relevance than state-of-the-art policies when it has enough budget, and AT-BSM shows better accuracy when changes are limited and K is small

	Measuring	Varying	AT-TSM	AT-BSM	WSJ-LRU
Hp.4.1	Relevancy	Refresh budget	B>3		
Hp.4.1	Accuracy	Refresh budget		✓	
Hp.4.2	Relevancy	CH	✓		
Hp.4.2	Accuracy	CH		✓	
Hp.4.3	Relevancy	K	✓		
Hp.4.3	Accuracy	K		K<7	K≥10
Hp.4.4	Relevancy	N	✓		
Hp.4.4	Accuracy	N		✓	
Overall	Relevancy	B >3	✓		
Overall	Accuracy	K<7		✓	

In general, the maximum improvement is the best achievement, and the minimum values are related to extreme conditions. They have no practical meanings, but show the worst conditions that are tested. In most of the cases, the best achievement is more than 10%. Among all the experiments, around half of the minimum improvements are positive, which shows that in half of the extreme cases, the proposed policy outperforms the ACQUA ones. The negative cases are not remarkable, but for experiment 3. In this experiment WSJ-LRU outperforms both proposed policies. This was expected since K is as large as the entire result set, and there is no advantage in focusing at the top.

For instance in Experiment 1, which there are various refresh budgets, considering $nDCG@k$, the maximum improvement of AT-TSM policy comparing to the WSJ-WBM policy is 11.68%, while the minimum improvement is -3.58%. The maximum improvement is related to the extreme condition of refresh budget 15, which is the best achievement, and the minimum is related to the refresh budget 1, in which WSJ-WBM policy performs better than AT-TSM. Refresh budget equal to 1 is an extreme case in which it is not possible to generate accurate results since there is not enough budget to update all the changes (see Fig. 6.9a).

Table 6.8 summarises the study of the research question for the four hypotheses presented in Sect. 6.5.3 and tested in the follow-up sections for different values of refresh budgets, CH, K, and N.

The results of Experiment 1 about Hp.4.1 show that, if there is enough refresh budget comparing to the K value, AT-TSM policy is the best option considering relevancy, while AT-BSM outperforms others when accuracy is more important.

The results of Experiment 2 about Hp.4.2 show that, for different values of change frequency CH, AT-TSM policy outperforms others in terms of relevancy, while AT-BSM generates more accurate top-k results, in terms of accuracy.

The results of Experiment 3 about Hp.4.3 show that, for different values of K, AT-TSM and AT-BSM perform better than others and the results are more relevant. However, considering accuracy, for low values of K, (i.e., $K < 7$), AT-BSM performs better than others, but for high values of K, ($K \geq 7$), WSJ-LRU is the best policy.

Finally, the results of Experiment 4 about Hp.4.4 show that AT-TSM and AT-BSM policies perform better than others. AT-TSM policy has a higher accurate result considering $nDCG@k$, while AT-BSM generates more accurate results considering $precision@k$. The results also show that giving more refresh budgets, it is possible to fill the gap between AT-TSM/AT-BSM and AT-ASM, and generates more relevant and accurate results.

Overall, AT-TSM shows better relevancy than state-of-the-art policies when it has enough budget and using $nDCG@k$ metric. AT-BSM shows better accuracy when changes are limited and K is small and $precision@k$ is measured.

6.6 Related Work

This study is the first to explore the evaluation of top-k continuous query for processing streaming and distributed data when the latter slowly evolves. Works near to this topic are in the domain of top-k query answering, continuous top-k query evaluation over streaming data, data sources replication, and federated query answering in RSP engine.

The top-k query answering problem has been studied in the database community, but none of the works in this domain has the focus of this work.

Ilyas et al. [7] present the generation of top-k result based on join over relations. Then, in [10] they extend relational algebra with ranking. Instead of the naïve materialize–then–sort schema, they introduce the rank operator. They extend relational algebra operators to process ranked list and they show the possibility to interleave ranking and processing to incrementally generate the ordered results. For a survey on top-k query processing techniques in relational databases see [8].

Yi et al. [20] introduced an approach to incrementally maintain materialized top-k views. The idea is to consider $top\text{-}k'$ results where k' is between k and parameter $Kmax$, to reduce the frequency of re-computation of top-k result which is an expensive operation.

There are also some initial works on top-k query answering in the Semantic Web community [11, 13, 17, 18] (see also Sect. 2.4).

Continuous top-k query evaluation has also been studied in the literature recently. All the works process top-k queries over data streams, but do not take into account joining streaming with distributed data, especially while they slowly evolve.

Mouratidis et al. [14] propose two techniques to monitor continuous top-k query over streaming data. The first one, the TMA algorithm, computes the new answer when some of the current top-k results expire. The second one, SMA, is a k-skyband based algorithm. It partially precomputes the future changes in the result in order to reduce the recomputation of top-k result. SMA has better execution time than TMA, but it needs higher space for "skyband structure" that keeps more than k objects.

As mentioned in Sect. 2.5, Yang et al. [19] were first in proposing an optimal algorithm in both CPU and memory utilization for continuous top-k query monitoring over streaming data.

There are also some works that evaluate queries over incomplete data streams like [6], or proposed probabilistic top-k query answering like [9].

Pripuzic et al. [15] also propose a probabilistic k-skyband data structure that stores the objects from the stream, which have high probability to become top-k objects in the future. The proposed data structure uses the memory space efficiently, while the maintenance process improves runtime performance compared to k-skyband maintenance [14].

Lv et al. [12] address the problem of distributed continuous top-k query answering. The solution splits the data streams across multiple distributed nodes and proposes a novel algorithm that extremely reduces the communication cost across the nodes. The authors call monitoring nodes those that process the streams and coordinator node the one that tracks the global top-k result. The coordinator assigns constraints to each monitoring node. When local constraints are violated at some monitoring nodes, the coordinator node is notified and it tries to resolve the violations through partial or global actions.

Zhu et al. [22] introduce a new approach that is less sensitive to the query parameters, and distributions of the objects' scores. Authors propose a novel self-adaptive partition-based framework, named SAP, which employs partition techniques to organize objects in the window. They also introduce the dynamic partition algorithm which enables SAP framework to adjust the partition size based on different query parameters and data distributions.

6.7 Outlook

This work studies the problem of continuously evaluating top-k join of streaming and evolving distributed data.

Monitoring top-k query over streaming data has been studied in recent years. Yang et al. [19] propose an optimal approach both in CPU and memory consumption to monitor top-k queries over streaming data. This study extends the approach focusing on joining the data stream with an evolving distributed data. Super-MTK+N *data structure* is introduced, which keeps the necessary and sufficient objects for top-k query evaluation, and handles slowly changes in the distributed data, while minimizing the memory usage.

The first proposed solution assumes that the engine can get notifications for all changes in the distributed data, and considers them as indistinct arrivals with new scores. This is often impossible over the Web, but it is interesting to explore the theoretical guarantees that the algorithm gives in terms of correctness in current and future windows. Topk+N *algorithm* is introduced, in which top-k results are affected and changed between two consecutive evaluations, based on the changes in the distributed data.

In order to guarantee reactiveness, the architectural approach of ACQUA keeps a replica of the distributed data and uses several maintenance policies to refresh such a replica.

This study, as a second solution, builds on such an architectural approach, and introduces AcquaTop *algorithm* that keeps up-to-date a local replica of the distributed data using alternatively AT-BSM *or* AT-TSM *maintenance policies*. AT-BSM policy maximizes the accuracy of the top-k result, and tries to get all the top-k answers. AT-TSM policy, instead, maximize the relevancy by minimizing the difference with the correct order, ignoring the accuracy of the results in lower positions.

To study the research question, four hypotheses are formulated, which test if the proposed policies provide better or at least the same accuracy (relevancy) comparing to the ACQUA policies for different parameters. The results of experiments show that AT-TSM policy has better relevance comparing to the ACQUA policies when it has enough budget. AT-BSM policy shows better accuracy when changes are limited and K value is small.

References

1. Balduini M, Della Valle E, Azzi M, Larcher R, Antonelli F, Ciuccarelli P (2015) Citysensing: fusing city data for visual storytelling. IEEE MultiMedia 22(3):44–53
2. Dehghanzadeh S, Dell'Aglio D, Gao S, Della Valle E, Mileo A, Bernstein A (2015) Approximate continuous query answering over streams and dynamic linked data sets. In: 15th international conference on web engineering, Switzerland
3. Della Valle E, Dell'Aglio D, Margara A (2016) Taming velocity and variety simultaneously in big data with stream reasoning: tutorial. In: Proceedings of the 10th ACM international conference on distributed and event-based systems. ACM, pp 394–401
4. Dell'Aglio D, Calbimonte J-P, Della Valle E, Corcho O (2015) Towards a unified language for rdf stream query processing. In: International semantic web conference. Springer, pp 353–363
5. Dell'Aglio D, Della Valle E, Calbimonte J-P, Corcho Ó (2014) RSP-QL semantics: a unifying query model to explain heterogeneity of RDF stream processing systems. Int J Semantic Web Inf Syst 10(4):17–44
6. Haghani P, Michel S, Aberer K (2009) Evaluating top-k queries over incomplete data streams. In: Proceedings of the 18th ACM conference on Information and knowledge management. ACM, pp 877–886
7. Ilyas IF, Aref WG, Elmagarmid AK (2002) Joining ranked inputs in practice. In: Proceedings of the 28th international conference on Very Large Data Bases, pp 950–961. VLDB Endowment
8. Ilyas IF, Beskales G, Soliman MA (2008) A survey of top-k query processing techniques in relational database systems. ACM Comput Surv (CSUR) 40(4):11
9. Jin C, Yi K, Chen L, Yu JX, Lin X (2008) Sliding-window top-k queries on uncertain streams. Proce VLDB Endow 1(1):301–312
10. Li C, Chang KC-C, Ilyas IF, Song S (2005) Ranksql: query algebra and optimization for relational top-k queries. In: Proceedings of the 2005 ACM SIGMOD international conference on Management of data. ACM, pp 131–142
11. Lopes N, Polleres A, Straccia U, Zimmermann A (2010) Anql: sparqling up annotated rdfs. Semant Web-ISWC 2010:518–533

12. Lv Z, Chen B, Yu X (2017) Sliding window top-k monitoring over distributed data streams. In: Asia-pacific web (APWeb) and web-age information management (WAIM) joint conference on web and big data. Springer, pp 527–540
13. Magliacane S, Bozzon A, Della Valle E (2012) Efficient execution of top-k sparql queries. Semant Web–ISWC 2012, pp 344–360
14. Mouratidis K, Bakiras S, Papadias D (2006) Continuous monitoring of top-k queries over sliding windows. In: Proceedings of the 2006 ACM SIGMOD international conference on management of data. ACM, pp 635–646
15. Pripužić K, Žarko IP, Aberer K (2015) Time-and space-efficient sliding window top-k query processing. ACM Trans Database Syst (TODS) 40(1):1
16. Seaborne A, Polleres A, Feigenbaum L, Williams GT (2013) Sparql 1.1 federated query. https://www.w3.org/TR/sparql11-federated-query/
17. Wagner A, Bicer V, Tran T (2014) Pay-as-you-go approximate join top-k processing for the web of data. In: European semantic web conference. Springer, pp 130–145
18. Wagner A, Duc TT, Ladwig G, Harth A, Studer R (2012) Top-k linked data query processing. In: Extended semantic web conference. Springer, pp 56–71
19. Yang D, Shastri A, Rundensteiner EA, Ward MO (2011) An optimal strategy for monitoring top-k queries in streaming windows. In: Proceedings of the 14th international conference on extending database technology. ACM, pp 57–68
20. Yi K, Yu H, Yang J, Xia G, Chen Y (2003) Efficient maintenance of materialized top-k views, pp189–200
21. Zahmatkesh S, Della Valle E, Dell'Aglio D (2016) When a filter makes the difference in continuously answering sparql queries on streaming and quasi-static linked data. In: International conference on web engineering. Springer, pp 299–316
22. Zhu R, Wang B, Yang X, Zheng B, Wang G (2017) Sap: improving continuous top-k queries over streaming data. IEEE Trans Knowl Data Eng 29(6):1310–1328

Chapter 7
Conclusion

Abstract This chapter reviews the investigation on query evaluation over streaming and evolving distributed data. This study focuses on the context of RSP engine, as it is an adequate framework to study continuous query answering over streaming and distributed data. Keeping a replica of distributed data, several maintenance policies are proposed for three classes of continuous queries that join streaming and distributed data: (*i*) queries that only contain 1:1 join relationship, (*ii*) queries that contain a filter constraint, and (*iii*) top-k queries.

Keywords Continuous query answering · Top-k query · Streaming data · Evolving distributed data · RDF data · RSP Engine

In this study, in order to attack the problem of query evaluation over streaming and evolving distributed data, the following research question was investigated:

> **RQ.** *Is it possible to optimize query evaluation in order to continuously obtain the most relevant combinations of streaming and evolving distributed data, while guaranteeing the reactiveness of the engine?*

This study focuses on the context of RSP engine, as it is an adequate framework to study continuous query answering over streaming and distributed data. Keeping a replica of distributed data, several maintenance policies are proposed for three classes of continuous queries that join streaming and distributed data: (*i*) queries that only contain 1:1 join relationship, (*ii*) queries that in addition contain a FILTER clause, and (*iii*) top-k queries over a 1:1 join relationship.

Chapter 3 studied the following sub-research question: *Given a query that joins streaming data with distributed data, how a local replica of distributed data can be refreshed in order to guarantee reactiveness while maximizing the accuracy of the continuous answer?*

This chapter proposed an approximated query answering over streaming and distributed datasets. In order to guarantee the reactiveness of the RSP engine, the architectural approach (ACQUA) is proposed, which uses a replica to store the distributed data at query registration time. Using refresh budget, which limits the number of access to the distributed data, can guarantee the reactiveness of the system. WSJ method, and WBM maintenance policy are proposed to keep the data in the replica fresh, so to provide more accurate answers.

Chapter 4 studied the following sub-research question: *Given a query that joins streaming data, returned from a WINDOW clause, with filtered distributed data, returned from a SERVICE clause, how a local replica of the distributed data can be refreshed in order to guarantee reactiveness while maximizing the freshness of the mappings in the replica?* The key difference between Chaps. 3, and 4 research questions is the FILTER clause.

Following the architectural approach of ACQUA, various maintenance policies are proposed to refresh the local replica. In the first step, the Filter Update Policy was proposed, which focuses on a band around the filtering threshold for selecting the mappings to update. Then, ACQUA.F Policies were introduced as a combination of the Filter Update Policy with ACQUA policies. Assuming that determining a band around the filtering threshold is straightforward, first, Filter Update Policy selects a set of mappings around threshold and, then, ACQUA policies process to the reduced set. The result of experiments showed that (*i*) Filter Update Policy outperforms ACQUA policies when the selectivity of filtering condition is above 60% of the total, and (*ii*) the combined policies keep the replica even fresher than the Filter Update Policy.

Further investigation of ACQUA.F approach showed that, in general, the assumption made in Chap. 4 does not hold: it is difficult to determine a priori the band around the filtering threshold to focus on. So, in Chap. 5, relaxing the assumption in the ACQUA.F policies, the rank aggregation approach was proposed, and the next sub-research question was explored: *Can rank aggregation be used to combine the ACQUA policies with* Filter Update Policy, *so to continuously answer queries (such as the one in Listing* 4.1) *and to guarantee reactiveness while keeping the replica fresh (i.e., giving results with high accuracy)?*

Chapter 5 proposed the rank aggregation approach, in which instead of applying in a pipe the Filter Update Policy and one of the ACQUA policies, each policy can express its *opinion* (by ranking data items according to its criterion) and, then, rank aggregation [5] takes fairly into account all opinions. The result of the experiments showed that the proposed policies are comparable to the ACQUA.F policies, but do not require to determine a priori the band to focus on.

Table 7.1 shows the summary of recommended policies based on the characteristic of data. The results show that for queries with 1:1 join relationship, WSJ-WBM outperforms others. For the class of queries containing FILTER clause, having to choose a policy, LRU.F$^+$ is the one that on average gives the best accuracy. For low selectivity ($<$60%), WBM.F* policy also generates accurate results comparable to LRU.F policy. Therefore, having the possibility to estimate the selectivity at run time, it would be better to use WBM.F* for low selectivities ($<$60%) and LRU.F$^+$ for high selectivities ($\geq$ 60%). So, proposing ACQUA.F$^+$ policies represented another significant step towards addressing the problem of getting the most relevant result in a timely fashion by evaluating query over streaming and distributed data.

Chapter 6 focused on continuous top-k query evaluation and investigated the next sub-research question: *How to optimize continuous, if needed approximate, top-k*

Table 7.1 Summary of recommended policies

Query features			Measuring	Conditions	Recommended policy
1:1 join	Filter	Top-k			
✓			Accuracy	any budget	WSJ-WBM
	✓		Accuracy	any selectivity or budget	LRU.F$^+$
	✓		Accuracy	selectivity <60%	WBM.F*
		✓	Relevancy	B >3	AT-TSM
		✓	Accuracy	CH <= 40, K <7	AT-BSM

queries that join streaming and distributed data, which may change between two consecutive evaluations, while guaranteeing the reactiveness of the system?

Although RSP-QL allows encoding top-k queries, the state-of-the-art RSP engines are not optimized for top-k queries and they would recompute the result from scratch at every evaluation as explained in [9, 11]. This recomputation bottleneck can lead RSP engines to lose their reactiveness. The state-of-the-art approach for top-k query evaluation [13] was extended, considering a distributed dataset that evolves.

The first solution, Topk+N algorithm, works in data centers where the infrastructure is under control. The data structure proposed in [13] was extended and the Super-MTK+N list was introduced. Then, the MinTopk algorithm [13] was modified by adding the ability of handling the indistinct arrival of objects, and considering changed objects as new arrivals.

The second solution focuses on evolving distributed data. Considering the architectural approach presented in Chap. 3 as a guideline, the AcquaTop framework was proposed. As ACQUA, it keeps a local replica of the distributed data but, differently, introduced two new policies tailored for top-k queries: AT-BSM, and AT-TSM. AT-BSM policy maximizes the accuracy of the top-k results, while AT-TSM policy maximizes the relevancy of the results.

The result of the conducted experiments showed that, when there is enough refresh budget, AT-TSM policy obtains more relevant results than any other policy proposed in this work. It was also found that AT-BSM policy generates more accurate results, when changes are limited and k is small (Table 7.1).

7.1 Suggestions for Future Works

This section discusses the limitations that are identified during the study and the possible extensions of the work as future directions.

First of all, this study focuses on the specific type of queries that contain a 1:1 join relationship between streaming and distributed data, and considers three classes of queries: (*i*) the one that contains only 1:1 join, (*ii*) the one that contains also a

FILTER clause, and (*iii*) top-k queries over 1:1 join relationship. As a future work, it is possible to broaden the class of queries which are subjects of the study:

- Queries with an 1:M, N:1, and N:M join relationship [7]. This types of query, requires to consider the selectivity of the join property in the maintenance policies. For example, for N:1 relationships, selecting an item in the SERVICE side with high selectivity of the join, can create many correct answers in the result.
- Queries that contain multi-join operators. Query optimization in this type of queries can be challenging. Different works tried to address the problem by proposing adaptive query processing such as [2] in database community and [1] in Semantic Web community.
- Preference queries that have qualitative formulation. This study works on top-k queries which are known as quantitative preference queries, where it is possible to formulate user's preferences as a scoring function. Preference queries [12], skyline queries [3], or top-k dominating queries [10] can be considered as an extensions of this work.
- Queries that contain other SPARQL clauses such as OPTIONAL, and UNION.
- Queries that contain multiple FILTER clauses or more complex filtering condition, e.g., having variables from WINDOW side in the FILTER clause.
- Top-k queries with text, meta-data, or hyper-textual searching.

The other limitation of this work is defining a static refresh budget to control the reactiveness of the RSP engine in each query evaluation. Further investigations can be done on dynamic use of refresh budgets following up ideas in [7], which proposed a flexible budget allocation method by saving the current budget for future evaluation, where it may produce better results.

Keeping the full replica of the distributed data is a feasible solution only for low volume datasets, which is one of the limitations in the proposed approach. For high volume distributed data, an alternative solution could be using a cache [4] instead of a replica, and considering recency or frequency strategies to keep the cache updated.

In the policies proposed in Chaps. 4, and 5, the work is limited to the combination of two policies. However, as another future work, it is possible to combine more than two maintenance policies and to explore how to dynamically determine the conditions for giving more priority to the specific policy through changing the weight related to each policy, e.g., by using the percentage of mappings subject to the filtering condition.

In the algorithm proposed in Chap. 6, a minimum threshold $min.score_S$ is defined in order to compute the new score for the changed objects that do not exist in Super-MTK+N list. As a future work, the approximation of the new score for this group of objects can be improved taking inspiration from [6].

This work considers a single stream of data and evaluates only one query in the experiments. However, more complex scenarios can be examined such as distributed streams and multiple queries. In distributed streams, it is needed to identify more efficient ways of communication and coordination between various nodes. In multiple queries scenario, while working on maximizing the relevancy of each query, it is worth to pay attention to the maintenance choices that bring overall benefit in the long term.

In this study, it is assumed that streaming and distributed data are complete. However, in the real world, there may be inaccurate or incomplete [8] data. As a future work, probabilistic methods and approximation algorithms can be considered to address the problem.

Last but not least, this study formulates the technology concepts and performs the evaluation with an experimental proof of concept outside any RSP engine, but as a future work the proposed framework can be implemented in an existing RSP engine (i.e., the C-SPARQL engine). This will enable applied research in the RSP practitioner community.

References

1. Acosta M, Vidal M-E (2015) Networks of linked data eddies: an adaptive web query processing engine for rdf data. In: International semantic web conference. Springer, pp 111–127
2. Avnur R, Hellerstein JM (2000) Eddies: continuously adaptive query processing. In: ACM sigmod record, vol 29. ACM, pp 261–272
3. Borzsony S, Kossmann D, Stocker K (2001) The skyline operator. In: Proceedings of the 17th international conference on data engineering. IEEE, pp 421–430
4. Dehghanzadeh S (2017) Cache maintenance in federated query processing based on quality of service constraints. PhD thesis
5. Dwork C, Kumar R, Naor M, Sivakumar D (2001) Rank aggregation methods for the web. In: WWW. ACM, pp 613–622
6. Fagin R, Lotem A, Naor M (2003) Optimal aggregation algorithms for middleware. J Comput Syst Sci 66(4):614–656
7. Gao S, Dell'Aglio D, Dehghanzadeh S, Bernstein A, Della Valle E, Mileo A (2016) Planning ahead: stream-driven linked-data access under update-budget constraints. In: International semantic web conference. Springer, pp 252–270
8. Haghani P, Michel S, Aberer K (2009) Evaluating top-k queries over incomplete data streams. In: Proceedings of the 18th ACM conference on Information and knowledge management. ACM, pp 877–886
9. Mouratidis K, Bakiras S, Papadias D (2006) Continuous monitoring of top-k queries over sliding windows. In: Proceedings of the 2006 ACM SIGMOD international conference on management of data. ACM, pp 635–646
10. Papadias D, Tao Y, Greg F, Seeger B (2005) Progressive skyline computation in database systems. ACM Trans Database Syst (TODS) 30(1):41–82
11. Pripužić K, Žarko IP, Aberer K (2015) Time-and space-efficient sliding window top-k query processing. ACM Trans Database Syst (TODS) 40(1):1
12. Stefanidis K, Koutrika G, Pitoura E (2011) A survey on representation, composition and application of preferences in database systems. ACM Tran Database Syst (TODS) 36(3):19
13. Yang D, Shastri A, Rundensteiner EA, Ward MO (2011) An optimal strategy for monitoring top-k queries in streaming windows. In: Proceedings of the 14th international conference on extending database technology. ACM, pp 57–68

Index

© The Author(s), under exclusive license to Springer Nature Switzerland AG 2020
S. Zahmatkesh and E. Della Valle, *Relevant Query Answering over Streaming
and Distributed Data*, PoliMI SpringerBriefs,
https://doi.org/10.1007/978-3-030-38339-8

Printed in the United States
By Bookmasters